DICTIONARY
of
SCIENCE

DICTIONARY
of
SCIENCE

ALICE GOLDIE BSc
GENERAL EDITOR: KEITH WHITTLES

GEDDES&
GROSSET

First published 1992
© 1992 Geddes & Grosset Ltd
New Lanark, Scotland

Cover design by Cameron Graphics Ltd
Glasgow, Scotland

ISBN 1 85534 099 2

Printed and bound in Great Britain

Contents

Dictionary of Science	7
Appendix 1: Periodic Table	278
Appendix 2: Element Table	279
Appendix 3: Greek Alphabet	283
Appendix 4: The International System of Units (SI units)	284
Appendix 5: Geological Time Scale	286
Appendix 6: The Solar System	288

A

abbe refractometer a device for measuring the REFRACTIVE INDEX of liquids.

aberration in physics, the result when a lens does not produce a true image. In astronomy, the apparent positional change of a body.

abscissa the horizontal or x-co-ordinate in GEOMETRY that is the distance of the point from the vertical or y-axis. For example, a point with CARTESIAN CO-ORDINATES (2,-6) has an abscissa of 2.

absolute temperature a temperature measured on the KELVIN SCALE with respect to ABSOLUTE ZERO.

absolute zero the temperature at which the particles of matter have no thermodynamic energy, theoretically given the lowest value of -273.15°C (-459.67°F).

ABO blood group *see* **blood grouping**.

AC *see* **alternating current**.

acceleration the rate of change of velocity with time. It is usually denoted by the symbol a and has SI units of metres per second squared (ms^{-2}). The acceleration of an object can be calculated using DIFFERENTIATION or the equation:

a = F/M where F = force and M = mass.

Independent of their mass, all bodies falling freely

acceptor

under the force of GRAVITY do so with uniform acceleration, particularly if air resistance is negligible. Acceleration due to gravity is approximately 9.8 ms^{-2} (32.17ft), thus where air resistance is negligible, the velocity of a falling object increases by 9.8 ms^{-1} per second. On the other hand, an object shot straight upwards with a particular velocity will decelerate by 9.8 ms^{-1} every second until it reaches its maximum height after a specific period of time. When its maximum height is reached, the object that had been shot straight upwards (no horizontal motion whatsoever) will start to fall with an acceleration of 9.8 ms^{-2}.

acceptor an ATOM that accepts ELECTRONS in forming a bond.

accretion in geology, the addition of material to the edge of a continent. Similarly, when applied to a celestial body, e.g. a planet, the process of accumulation of matter onto the body, driven by the influence of GRAVITY.

acetylene *see* **ethyne**.

acid any substance that releases hydrogen ions during a chemical reaction, thus lowering the PH of the solution. An acidic solution has a pH of less than 7 and will react with a BASE to form a salt and water. An acid can be thought of as a proton (H$^+$) donor (BRÖNSTED-LOWRY definition) or as an electron pair acceptor (LEWIS definition).

acid rain rain with a high concentration of pollutants, such as dissolved sulphur and nitrogen oxides, which have harmful effects on plant and animal life. The pollutants are principally byproducts of industrial activities involving the burning of coal or oil (*see* FOSSIL FUELS).

acquired immune deficiency syndrome *see* **AIDS**.

acquired immunity *see* **adaptive immune system, immune system**.

actinides those elements, from actinium (Ac) to lawrencium (Lr), which mirror the former in their chemistry. All the elements beyond actinium in the PERIODIC TABLE are radioactive, and many of them do not occur naturally.

action potential a voltage pulse produced in a nerve by a stimulus. A continuing stimulus produces frequent pulses resulting in muscular responses.

activation energy the input of energy required to initiate a chemical reaction. In some reactions, the KINETIC ENERGY of the reactant molecules is sufficient for the reaction to proceed, but in others the activation energy is too high for the reaction to occur (*see* CATALYST).

acute an angle that has a size that lies between 0° and 90°.

adaptation a feature of an organism that has evolved under natural selection as it enables the organism to function efficiently in a specific aspect of its particular environmental NICHE. The adaptation can be either psychological, where exposure to certain environmental conditions will only cause a change in the organism's behaviour, or it can be genetic, as when the organism possesses GENES that produce characteristics that prove to be beneficial for survival in its environment. Examples of genetic adaptation are the genetic changes that occur in bacterial populations, conferring antibiotic resistance, as in the strains of bacteria resistant to penicillin. Such antibiotic resistance is a major problem in hospitals when trying to prevent any risk of post-operative infections.

adaptive immune system one functional division of the IMMUNE SYSTEM that produces a specific response to any PATHOGEN. There are two mechanisms of acquired or adaptive immunity:

(1) Humoral—the production of soluble factors, called ANTIBODIES, by activated B-CELLS.

(2) Cell-mediated immunity—the range of T-CELLS involved in the specific recognition of an antigen, which must be present on the surface of another cell, i.e. an antigen-presenting cell (APC). The immunity produced, **adaptive immunity**, is also known as **acquired immunity** because the adaptive immune system is capable of remembering any infectious agent that has induced the proliferation of B-cells (it has acquired a memory for that particular agent, *see* B-CELLS).

adaptive radiation the evolutionary separation of species into numerous descendent species in order to exploit the various habitats that exist throughout the world. This evolutionary divergence of species probably explains the bewildering array of, say, amphibians, which has arisen as a result of adaptive radiation after the first amphibians moved on to land.

addition formula an equation used to express the sum or difference of angles as a sum or difference of the products of the trigonometric functions of the individual angle. For example, the formulae for the functions of cosine and sine are

$$\text{cosine } (A+B) = \cos A \cos B + \sin A \sin B$$
$$\text{sine } (A+B) = \sin A \cos B + \cos A \sin B$$

The above formulae can be used to derive others for functions such as the tangent, $\cos^2 A$, and other trigonometric functions.

addition polymerization the formation of a large molecule called a POLYMER, the structure of which is a repeating array of atoms. The reaction involves the successive addition of molecules of the MONOMER to form a long chain. Addition polymerizations often occur at high temperatures and pressure in the presence of a CATALYST.

addition reaction a chemical reaction in which two parts of one reactant add to a multiple bond of the other reactant, one part to each side of the bond. An addition reaction will usually involve ALKENES, but at the end of the reaction they will no longer contain double bonds but instead have single bonds (the structure of an ALKANE).

adenine ($C_5H_5N_5$) a nitrogenous base component of the NUCLEIC ACIDS, DNA and RNA, which has a PURINE

adenosine triphosphate

structure. In DNA, adenine will always base pair with THYMINE, but in RNA, during TRANSCRIPTION, adenine will base pair with URACIL. Adenine is also present in other important molecules, such as ATP.

adenosine triphosphate *see* **ATP**.

adiabatic change a change that occurs with no change in heat content, thus if the volume and pressure of an enclosure alter without heat gain from or heat loss to the surroundings, the change is adiabatic.

ADP *see* **ATP**

adrenal gland *see* **endocrine system**.

adsorption the taking up, or concentration of, one substance at the surface of another, e.g. a dissolved substance on the surface of a solid. Chemisorption is when COVALENT bonds hold a layer of atoms or molecules of the dissolved substance on the surface.

aerobic respiration a set of enzyme-catalysed reactions requiring oxygen to release energy during the degradation of glucose. The first stage of aerobic respiration, GLYCOLYSIS, occurs in the CYTOPLASM of both plant and animal cells. However, the remaining two stages occur in the MITOCHONDRIA of EUCARYOTIC cells and the membrane of PROCARYOTIC cells. Thirty-eight molecules of ATP are generated per molecule of glucose undergoing complete aerobic respiration.

aerosol a fine mist or fog in which the medium of dispersion is a gas. Also, a pressurized container with spray mechanism used extensively for deodorants, insecticides, etc.

AIDS (*acronym for* Acquired Immune Deficiency Syndrome) a serious disease thought to be caused by a human retrovirus called HIV—human immune deficiency virus—which kills the adult T-CELLS of the IMMUNE SYSTEM. This destruction of a critical aspect of the body's immune system leaves the patient open to both minor and major infections, as well as the possibility of developing cancer. Not all HIV-infected individuals develop AIDS, but it can be passed on by the following methods: the receiving of infected blood during transfusions; a mother passing it on to a foetus; the sharing of needles in drug abuse; and the passing of body fluids during sexual contact. There is no evidence that HIV can be transmitted by everyday activities and social contact, such as swimming, sharing cutlery, using a public toilet, etc.

alcohol an organic compound similar in structure to an ALKANE but with the FUNCTIONAL GROUP -OH (hydroxyl) instead of a hydrogen atom. Alcoholic beverages contain ETHANOL, the alcohol obtained during the fermentation of sugars or starches.

aldehyde (*also known as* **alkanal**) any of a class of organic compound with the CO-RADICAL attached to both a hydrogen atom and a hydrocarbon (alkyl) radical, giving the type formula R.CO.H.

alga (*plural* **algae**) the common name for a simple water plant, which is without root, stem or leaves but which contains CHLOROPHYLL. Algae range in form from single cells to plants many metres in length. The

algebra

blue-green algae, cyanobacteria, are widely distributed in many environments.

algebra the use of symbols, particularly letters, in solving mathematical problems in order to find the value of an unknown quantity or to study complex relationships, systems and theories. For example, the physicist, Albert EINSTEIN, used advanced algebra to derive equations from his general theory of RELATIVITY.

algorithm a set of rules comprising a mathematical procedure that enables a problem to be solved in a specific number of steps. Each rule is precise in its definition so that, theoretically, the process can be undertaken by a machine.

alkali a soluble base that will give its solution a pH value greater than 7. The hydroxides of the metallic elements sodium (Na) and potassium (K) are strong alkalis, as is ammonia solution (NH_4OH).

alkali metals the metals lithium (Li), sodium (Na), potassium (K), rubidium (Rb), and caesium (Cs), which belong to group 1A of the PERIODIC TABLE. The last member of the group, francium (Fr), occurs only as a radioactive isotope. All the metals have one electron in their outer shell and are thus univalent. The elements are all prepared by ELECTROLYSIS of the fused HALIDES. Their melting and boiling points fall with increasing atomic weight, so that caesium has the second lowest melting point of any metal.

alkaline earth metals the metals of group 2A of the

PERIODIC TABLE: beryllium (Be), magnesium (Mg), calcium (Ca), strontium (Sr), barium (Ba) and radium (Ra). These bivalent (valency of two) elements have properties similar to the ALKALI METALS but are, in general, less volatile.

alkaloids basic organic substances found in plants and with a ring structure and the general properties of amines. In nature, they act as deterrents for the plant against herbivores. Many alkaloids are used in medicine, e.g. cocaine, codeine, MORPHINE, QUININE.

alkanal *see* aldehyde.

alkane an open-chain HYDROCARBON with single bonds between each carbon atom. The first member of this HOMOLOGOUS SERIES is METHANE, CH_4, and the subsequent members may be considered to be derived from methane by the simple addition of the unit, $-CH_2$. All the members conform to the general formula of C_nH_{2n+2}, thus the chemical formula for the second member of the series, ETHANE, is C_2H_6. Alkanes are SATURATED COMPOUNDS as all the valence electrons of the carbon atoms are engaged within the single, COVALENT BONDS. As the alkanes are saturated compounds, they are quite stable but will undergo a slow SUBSTITUTION reaction with a HALOGEN. For example:

CH_4 (g) + Cl_2 (g) ———> CH_3Cl (g) +   HCl (g)
methane chlorine   chloromethane hydrogen chloride

alkene an open-chain HYDROCARBON containing a carbon-carbon double bond. The first member of this HOMOLOGOUS SERIES is ETHENE, C_2H_4, and all other

alkyne

members conform to the general formula C_nH_{2n}. Alkenes are UNSATURATED hydrocarbons, which can easily undergo an ADDITION REACTION as each carbon atom has one available ELECTRON that is not engaged in the formation of the double bond.

alkyne an open-chain HYDROCARBON containing a carbon-carbon triple bond somewhere in its molecular structure. The first member of the alkynes is ETHYNE (also called acetylene), C_2H_2, and the general formula for the other members is C_nH_{2n-2}. They are UNSATURATED compounds that will readily undergo ADDITION REACTIONS across their triple bond, due to the availability of the four electrons engaged in the formation of two of the carbon-carbon bonds of the triple bond. The remaining carbon-carbon bond of the C=C bond and any carbon-hydrogen bond are a much stronger type of bond, which makes them more stable and less reactive.

allele one of several particular forms of a GENE at a given place (locus) on the CHROMOSOME. Alleles, responsible for certain characteristics of the PHENOTYPE, are usually present on different chromosomes. It is the dominance of one allele over another (or others) that will determine the phenotype of the individual.

allergy (*plural* **allergies**) an overreactive response of the immune system of the body to foreign ANTIGENS. Allergies result from the hyperproduction of one class of ANTIBODY, IgE, which activates the release of certain products, including histamine, by the MAST CELLS. This causes the characteristic symptoms of an allergy, i.e. inflammation, itching etc. An induced overreaction by the immune system can be produced by an environmental substance like pollen, certain foods, or even toxins injected by insects such as wasps. Such reactions can be counter-attacked using antihistamine drugs.

allotropy the state of an ELEMENT when it exists in two or more forms with different physical properties. Sulphur is the oft-quoted example.

alpha decay the emission of ALPHA PARTICLES from the nucleus of a radioactive ISOTOPE.

alphanumeric a set of characters derived from the numerals 0 to 9 and the alphabet. In computing, the remaining keyboard characters are used for functions other than keying text.

alpha particle a fast-moving helium nucleus ($^4_2\text{He}^{2+}$) with a short, straight range from the source of emission.

alternating current (AC) an electric current that reverses the direction of its flow at constant intervals of time, resulting in a constant frequency independent of the type of CIRCUIT. Mains electricity is comprised of alternating current as opposed to DIRECT CURRENT.

altitude in geometry, the line segment that measures the perpendicular distance from a vertex of a POLYGON to the opposite side of that polygon, e.g. the height of an isosceles triangle.

aluminium

aluminium a light, ductile and malleable metal, which is a good conductor of electricity. It is found in group 3A of the PERIODIC TABLE and is the most common metallic element in the earth's crust (third most common overall, at 8 per cent). It is extracted from the hydrated oxide (bauxite) by electrolysis with molten cryolite as a FLUX. The metal and its alloys are used for innumerable purposes, including the manufacture of cooking utensils, electrical equipment, aircraft, etc.

American Standard Code for Information Interchange *see* **ASCII**.

amines compounds formed from ammonia, NH_3, by the replacement of one or more hydrogen atoms with organic RADICALS. There are three classes: primary, NH_2R; secondary, NHR_2, and tertiary, NR_3.

amino acid any of the 20 standard organic compounds that serve as the building blocks of all PROTEINS. All have the same basic structure, containing an acidic carboxyl group (-COOH) and an amino group (-NH2), both bonded to the same central carbon atom referred to as the α-carbon. Their different chemical and physical properties result from one variable group, the side chain or R-group, which is also attached to the α-carbon.

amino group an essential part ($-NH_2$) of the AMINO ACIDS.

ammonia a COVALENT compound (NH_3) that exists as a colourless gas. It will react with water, giving the alkaline solution known as ammonium hydroxide

(NH₄OH). Ammonia will ionize in water to form the ammonium ion (NH_4^+) and the hydroxide ion (OH^-) $NH_3\ (g) + H_2O\ (l) \longrightarrow NH_4^+(aq) + OH^-(aq)$.

ampere (A) the unit for measuring the quantity of electric CURRENT, often abbreviated to **AMPS**.

amphoterism a property of the oxides or hydroxides of certain metallic elements, which allows them to function both as acids and alkalis. Although insoluble in water, amphoteric compounds will dissolve in acidic solutions (pH < 7) or basic solutions (pH > 7).

amplitude the maximum displacement of a particle from its position of rest. For sound waves, the amplitude relates to the intensity of the wave and is the distance between the X-AXIS (rest position) and the crest or trough of the wave. In a simple pendulum, the amplitude is defined as the angle through which the arm moves when swinging between its extreme and rest positions.

AMPS *see* ampere.

anabolism the biosynthetic building-up processes of METABOLISM. For example, the synthesis of fatty acids and phospholipids in the GOLGI APPARATUS.

anaemia *see* erythrocyte.

anaerobic respiration a set of enzyme-catalysed reactions releasing energy from the SUBSTRATE in the absence of oxygen. The first stage of anaerobic respiration is GLYCOLYSIS—as in AEROBIC RESPIRATION—but there is only one other stage, FERMENTATION, which has two alternative pathways as follows:

anaphase

(1) The production of lactic acid, as in muscle cells during prolonged contraction.
(2) The production of the alcohol, ethanol, by micro-organisms such as yeast or some plants.

In all anaerobic respiration, only two ATP molecules are generated (during glycolysis) per glucose molecule.

anaphase *see* **mitosis.**

anion a negatively charged ion formed by an atom or group of atoms gaining one or more electrons.

anisotropy the state of a substance when it possesses a property dependent upon direction, e.g. some crystals have a different REFRACTIVE INDEX in different directions. The opposite of ISOTROPY.

annealing the process whereby materials, usually metals, are heated to, and held at, a specific temperature before controlled cooling. This relieves STRAIN set up by other processes.

anode the positively charged electrode of an electrochemical cell to which the ANIONS of the solution move to give up their extra electrons, i.e. the electrode at which OXIDATION occurs.

antibiotic a chemical produced by micro-organisms, such as BACTERIA and moulds, that can kill bacteria or prevent their growth. The first antibiotic to be discovered was penicillin, and there are now many more, including erythromycin, streptomycin and terramycin.

antibody (*plural* **antibodies**) a protein circulating in

antibody

the blood, which is produced by B-CELLS and will bind to the surface of an ANTIGEN. The production of antibodies is a specific immune response of the ADAPTIVE IMMUNE SYSTEM. Antibodies consist of protein chains that form IMMUNOGLOBIN, i.e. Ig, and are very useful for identifying specific types of protein unique to a particular plant or animal or to a VIRUS that may be circulating throughout the body of an individual. Although millions of different antibodies are produced in order to cope with any PATHOGEN that may arise, there are only five major classes, which have the following functions:

Antibody Function

Antibody	Function
IgG	The most abundant, it combats microorganisms and toxins. As it can cross the placenta, it is the first Ig found in newly born infants.
IgA	The major Ig in mucosal secretion, it defends external surfaces including the gut wall to help cope with antigens found within the gut.
IgM	The first Ig to be produced during infection, it is very effective against bacterial infections.
IgD	This is present on surfaces of B-cells, but no specific function is known for it.
IgE	This protects external surfaces and triggers the release of histamine from other cells of the immune system.

antigen any substance that triggers an immune response due to the body's IMMUNE SYSTEM recognizing it as foreign. Common antigens are proteins present on the surface of bacteria and viruses. Unsuccessful transplant operations are usually a result of the patient's immune response recognizing the surface cells of the organ from the donor as non-self. The organ is said to be rejected when the patient's immune system becomes activated and tries to destroy the donated organ.

antimatter matter, as yet hypothetical, which is composed of **antiparticles**, i.e. particles of the same mass but opposite values for its other properties such as charge. The ELECTRON and POSITRON are particle and antiparticle, and interaction between them would result in annihilation and the release of energy.

aorta the largest ARTERY in the body, which acts as the blood outflow of the left VENTRICLE of the heart. The aorta has an approximate diameter of three centimetres, with thick muscular walls to carry blood under pressure. The aorta divides into several branches, which supply blood to the arms and the head. It then continues down around the spine to the level of the lower abdomen, where it divides into two major branches to supply the legs.

APC *abbreviation for* antigen-presenting cell. *See* ADAPTIVE IMMUNE SYSTEM.

apogee the point at which a satellite is at its greatest distance from the earth. The converse situation is called the perigee.

aquifer a permeable ROCK, underlain by impermeable strata, which contains significant quantities of recoverable water.

Archimedes' principle a law of physics stating that when a body is partly or totally immersed in a liquid, the apparent loss in weight equals the weight of the displaced liquid.

arithmetic series the sum of the terms of a sequence of quantities. An unknown term (nth term) can be calculated using the formula below:

$$n^{th} \text{ term} = a + (n - 1)d$$

The sum of n terms is calculated using:

$$S_n = n/2 \, [2a + (n - 1)d]$$

where "a" is the first term and "d" is the common difference of the series.

armature the rotating coil of an electric motor. More generally, it is any electric component within a piece of equipment in which a VOLTAGE is induced by a MAGNETIC FIELD.

aromatic hydrocarbon any compound that has a molecular structure based on that of BENZENE. Aromatic hydrocarbons are UNSATURATED, closed-chain compounds that will undergo SUBSTITUTION reactions as well as ADDITION REACTIONS, depending on which functional groups are present within their structure and the reactivity of the other reagents.

arteriosclerosis the thickening, hardening and loss of elasticity of the ARTERIES. This can be a pathological

artery

condition of advancing age, or it can be associated with fatty deposits, particularly CHOLESTEROL, that block the arteries, causing their diameter to decrease. As a result, the heart must strain to increase its muscular activity to generate enough pressure to pump the blood through the arteries.

artery (*plural* **arteries**) a thick-walled vessel that carries blood under pressure resulting from the pumping mechanism of the heart. The PULMONARY ARTERY carries deoxygenated blood from the heart to the lungs, but all other arteries carry oxygenated blood from the heart to body tissues.

ASCII (*acronym for* American Standard Code for Information Interchange) a standard code of 128 alphanumeric characters for storing and exchanging information between computer programs.

ascorbic acid another term for the water-soluble vitamin C found in all citrus fruits and green vegetables (especially peppers). A deficiency of ascorbic acid leads to the fragility of tendons, blood vessels and skin, all of which are characteristic of the disease called scurvy. The prescence of ascorbic acid is also believed to help in the uptake of iron during the process of digestion by the body.

assay the chemical analysis of a mixture to determine the amount of a particular constituent, e.g. metal in an ore.

asteroid one of many rocky or metallic bodies, more correctly called planetoids, that orbit the sun between the orbits of Mars and Jupiter. Most are very small, but the largest, Ceres, has a diameter of about 1,000 kilometres. Meteorites are debris from the asteroid belt formed by the collision of the bodies.

asymptote a straight line that a curve of a FUNCTION approaches but never reaches. Although the perpendicular distance between the curve and its asymptote will decrease and eventually equal zero, the distance between the origin and the asymptote will tend to infinity, and there will still be no contact between the curve and the asymptote.

atmosphere (1) a unit of PRESSURE defined as the pressure that will support a mercury column 760mm high at 0°C, sea level, and a latitude of 45°. (2) the layer of gases surrounding the earth, which contains, on average, 78 per cent nitrogen, 21 per cent oxygen, almost 1 per cent argon, and then very small quantities of carbon dioxide, neon, helium, krypton and xenon. In addition, air usually contains water vapour, hydrocarbons and traces of other materials and compounds.

atom the smallest particle that makes up all matter and still retains the chemical properties of the element. Atoms consist of a minute nucleus containing PROTONS (p) and NEUTRONS (n), with negatively charged particles called ELECTRONS (e) moving around the nucleus. The number of protons in an atom is equal to the number of electrons, and as the protons are positively charged, the atom is, overall, electrically neutral (*compare* ISOTOPE). The various elements of the

atom bomb *see* **nuclear fission**.

atomic mass unit (*see* RELATIVE ATOMIC MASS) defined in 1961 as one twelfth of the mass of an atom of ^{12}C, having formerly been based upon ^{16}O, the most abundant ISOTOPE of oxygen.

atomic number (A, at. no.) the number of protons in the NUCLEUS of an ATOM. Although all atoms of the same element will have the same number of protons, they can differ in their number of neutrons, resulting in an ISOTOPE.

atomic weight *see* **relative atomic mass**.

ATP (*abbreviation for* adenosine triphosphate) an important molecule that is used as energy to drive all cellular processes. It consists of ADENINE and a 5-ring sugar that has three phosphate (PO_4) groups attached by high-energy bonds. ATP can be synthesized during GLYCOLYSIS by addition of a phosphate group to adenosine diphosphate (ADP), or it can be broken down to form the ADP, releasing energy that will be used to drive a metabolic process, such as active transport across cell membranes or the contraction of muscle cells.

atrium (*plural* **atria**) a minor chamber of the heart that is considered to be a reservoir as blood passes from it into the pumping chamber, the VENTRICLE. The right atrium of the heart receives the blood carried by the superior and inferior VENA CAVA before it passes via a valve into the right ventricle. The pulmonary veins carry oxygenated blood from the lungs into the left atrium, which then flows via a valve into the left ventricle.

aurora luminous, and often colourful, sheets or streaks in the sky, formed by high-speed, electrically charged solar particles entering the upper atmosphere where electrons are released, thus creating molecules with the associated release of light. These effects are related to SUNSPOT activity and are termed the Northern Lights (*aurora borealis*) in the northern hemisphere, and the Southern Lights (*aurora australis*) in the southern hemisphere.

autolysis *see* **lysis**.

autosome a biological term describing all CHROMOSOMES within a cell except the SEX CHROMOSOMES. In a DIPLOID cell, there are two copies of every autosome, each of which will carry genetic information for the same aspect of the individual's PHENOTYPE. Although autosomes are not involved with determining the sex of an individual, they can carry genetic information that will affect the sexual characteristics.

Avogadro's constant *or* **Avogadro's number** the number of particles present in one MOLE of a substance. It is given the symbol N or L and has the value of 6.023×10^{23}.

Avogadro's law the principle formulated by the Italian scientist Amedeo Avogadro (1776-1856) that states that equal volumes of all gases contain the same number of molecules when under the same temperature and pressure. For the purpose of calculation, one MOLE of gas will occupy a volume of 22.4 litres at standard conditions, i.e. 273.15K and 1 atmosphere.

axon *see* **neuron**.

B

backcross a mating experiment that is used to discover the GENOTYPE of an organism. It involves crossing the organism of unknown genotype with an organism of known genotype (usually the HOMOZYGOTE recessive). The PHENOTYPES of the produced progeny should directly correspond to the chromosomes of the parental organism of unknown genotype. A backcross usually reveals whether the unknown genotype is homozygous or heterozygous (*see* HETEROZYGOTE) for a particular gene.

bacterium (*plural* **bacteria**) a micro-organism, usually unicellular, which does not photosynthesize (*see* PHOTOSYNTHESIS) and which causes diseases that are treated with ANTIBIOTICS. Bacteria occur in water, air, soil and rotting animal or plant debris (SAPROTROPHS). They are classified by shape into three main groups: the spherical coccus form; the spiral spirillum; and the rod-shaped bacillus.

ballistics the study of the flight path of an object moving under the influence of gravity.

bar chart *or* **bar graph** a graph that illustrates relationships between variables by vertical parallel bars, with the heights of the bars representing the data

barometer proportionally.

barometer an instrument used to measure the pressure that the atmosphere exerts on the earth's surface, which helps to predict impending weather changes.

base in a chemical reaction, any substance that dissociates in water to produce hydroxide ions (OH$^-$). A base can be thought of as a proton (H$^+$) acceptor or as an electron pair donor. It will react with an acid to give a salt and water (the latter formed from the OH$^-$ ion from the base and the H$^+$ ion from the acid).

$$NaOH(aq) + HCl(aq) \longrightarrow NaCl(aq) + H_2O$$

In mathematics, a base is the number raised to a certain power (EXPONENT), which will produce a fixed number. For example, base 5 to the power of 3 will equal 125, i.e. $5^3 = 125$.

base pair the arrangement of two nitrogenous molecules on opposite strands of a DNA double-helix or a DNA-RNA molecule. The bases can be classified according to their structure: ADENINE and GUANINE are PURINES, whereas THYMINE, CYTOSINE and URACIL are PYRIMIDINES. As a consequence of geometrical factors, a purine will be hydrogen-bonded to a pyrimidine, i.e. A-T and C-G. This specific base-pairing keeps the DNA structure in a highly organized order, allowing the replication of DNA to be very precise, thus ensuring that each daughter cell will inherit the same genetic information contained within the parent cell.

BASIC (*acronym for* Beginners All-purpose Symbolic

Instruction Code) a simple language for procedural programming and interacting with a computer.

B-cells these are LYMPHOCYTES, which differentiate in the bone marrow to form part of the IMMUNE SYSTEM of humans. B-cells become activated when they encounter a specific ANTIGEN, leading to proliferation and secretion of ANTIBODIES by the activated B-cells. After a first encounter with an antigen, some of the B-cells remain and serve as memory cells. The memory cells will be capable of recognizing the same antigen during any subsequent encounter and will, therefore, produce a faster and greater secondary response (this is the principle behind vaccination).

Beaufort scale a system for indicating wind velocity, with measurements taken at 10 metres (32.8 ft) above ground level. The numerical scale ranges from 0 (calm, speed < 0.3 ms^{-1}) to 12 (hurricane, speed > 32.7 ms^{-1}).

bentonite a useful clay formed by the breakdown of volcanic ash, which has properties similar to FULLER'S EARTH. It is THIXOTROPIC and is used in the construction, paper and pharmaceutical industries and as an additive in oil-drilling muds.

benzene a toxic hydrocarbon that is a liquid at room temperature and has the chemical formula C_6H_6. The six carbon atoms of benzene form a ring, and the overall molecular structure is planar, with all bond angles having the same value of 120°. The benzene ring is a very stable structure due to the delocalization of six electrons (one electron is contributed by each carbon atom) and therefore does not readily undergo an ADDITION REACTION. All AROMATIC HYDROCARBONS contain a benzene ring within their molecular structures.

beri-beri a crippling human disease that is caused by dietary deficiency of the water-soluble vitamin B, also called thiamine, which is essential for the metabolic conversion of carbohydrate to glucose. Affected individuals show symptoms of muscle atrophy, paralysis, and mental confusion, and may eventually suffer from heart failure. Chronic alcoholism will lead to beri-beri, and this is thought to be responsible for the development of a psychosis called KORSAKOFF'S SYNDROME.

beta decay a type of ionizing radiation that emits either a negatively charged ß-particle (e^-, an electron) or a positively charged ß-particle (e^+, a positron). In electron emission, the PROTON number of the nucleus increases by one, whereas in positron emission, the proton number decreases by one. Both types of ß-particles tend to have a longer range than those emitted during ALPHA DECAY and produce a scattered rather than a straight path through matter.

bicarbonates acid salts of carbonic acid (H_2CO_3), where one of the hydrogens has been replaced by a metal. Aqueous solutions contain the bicarbonate ion (HCO_3^-).

bile a viscous fluid produced by the liver and stored in a small organ, the gall bladder, near the liver. Bile is an alkaline solution consisting of bile salts, bile pigments and CHOLESTEROL, which aids in the digestion of fatty particles present in the diet. Food entering the DUODENUM triggers the muscular contraction of the gall bladder wall, and bile is forced into the duodenum via the bile duct. Although bile does not contain any digestive enzymes, the bile salts help in the digestion of fatty food particles by acting as emulsifying agents, i.e. they break down large fat particles into many smaller ones, a process that exposes a larger surface area to the digestive action of the enzymes, LIPASES.

binary digit *see* **bit**.

binary star two stars that are mutually attracted by gravitational forces, thus forming a double star where both bodies revolve around a common centre of mass.

binary system an arithmetical code that uses a combination of the two digits, 1 and 0, expressed to base 2. The value of any digit in a binary number increases by powers of 2 with each move to the left. For example, 1010.1 to base 2 in the binary system (written 1010.1^2) represents:

$(1 \times 2^3) + (0 \times 2^2) + (1 \times 2^1) + (0 \times 2^0) + (1 \times 2^{-1})$

which, in the everyday decimal system, adds up to:

$8 + 0 + 2 + 0 + \frac{1}{2} = 10\frac{1}{2}$.

The binary number system is the basis of the internal coding in all modern computing, as the two digits, 0 and 1, are represented as the on/off states of any switch in a circuit.

binomial theorem a mathematical formula whereby the expansion of any positive power of a binomial (a + b) to a POLYNOMIAL may be reached without performing the numerous multiplications:

$(a + b)^n = a^n + n a^{n-1} b + n(n - 1)a^{n-2}b^2/2! + ... + b^n$.

For example, the binomial $(a + b)^5$ has the following expansion:

$(a + b)^5 = a^5 + 5a^4b + 10a^3b^2 + 10a^2b^3 + 5ab^4 + b^5$

binomics *see* **ecology**.

biochemistry a method for investigating the chemistry of the biological processes occurring in the cells of all organisms. Such investigations provide an understanding of a broad range of important processes, from the control of cell metabolism to the effects that a certain disease has upon the cells of the body.

biological (*or* **biochemical**) **oxygen demand** (BOD) a measure of the pollution of effluent where microorganisms take up dissolved oxygen in decomposing the organic material present in the effluent. BOD is quantified as the amount of oxygen, in milligrams, used by one litre of sample that is stored in the dark at 20°C for five days.

biosphere the region of the earth's surface (both land and water) and its immediate atmosphere, which can support any living organism.

biosynthesis the production of complex chemical compounds by living organisms using ENZYMES.

biotechnology the industrial use of organisms, their parts or processes, to produce drugs, food or other useful products. Modern processes include the controlled growth of specific fungi in laboratories to obtain the antibiotic, penicillin, and the production of alcohol during FERMENTATION in yeast. The scope for biotechnology is enormous, with a great deal of research being directed towards GENETIC ENGINEERING in plants and animals.

bit (*abbreviation for* **binary digit**) this is either the number 1 or 0. A bit is the smallest unit of a computer capable of storing one unit of information, although the storage capacity of the computer is measured in BYTES.

black body a body that absorbs all heat or light radiation falling upon it. In practice, no body can achieve this state.

black hole a region in space from which matter and radiation cannot escape due to the intense gravitational field. Their origin is thought to lie with the collapse of massive stars. Since no radiation is emitted, black holes cannot be observed directly. However, it is believed that visible stars form BINARY STARS with black holes and the capture of matter by the black hole from the visible star allows indirect observation, because X-rays are radiated as the matter falls into the black hole.

blood a vital substance consisting of red blood cells (ERYTHROCYTES) and white blood cells (LEUCOCYTES) suspended in a liquid medium called BLOOD PLASMA. As well as many proteins, mammalian blood also con-

blood clotting

tains small disc-shaped cells called platelets, which are involved in BLOOD CLOTTING. Blood circulates throughout the body and serves as a mechanism for transporting many substances. Some of the essential functions of blood are:

(1) Oxygenated blood is carried from the heart to all tissues by the arteries while the veins carry deoxygenated blood, which contains carbon dioxide, back to the heart.

(2) Essential nutrients, such as glucose, fats and AMINO ACIDS (the building blocks of proteins), enter the blood from the intestinal wall, or the liver and fatty deposits, and are carried to all the regions of the body.

(3) The metabolic waste products, ammonia and carbon dioxide, are carried to the liver, where they react to form urea, which is then carried by the blood to the kidneys for excretion.

(4) Steroid and thyroid hormones—important regulatory molecules—are carried to their target cells after they are secreted into the blood by the ENDOCRINE SYSTEM. Although insoluble molecules, they are carried in soluble particles called low density LIPOPROTEINS.

blood clotting (*also called* **haemostasis**) a process, involving many chemical factors, that stops blood leaking from an area of injured tissue. In the first instance, constriction of any blood vessels in the injured tissue restricts the leakage of blood, and the subsequent formation of a plug helps to seal off the damaged area, preventing the entry of micro-organisms. The formation of this plug is triggered by an enzyme secreted by damaged blood vessels and blood platelets and is completed after a chain of chemical reactions. The final hard clot consists of blood platelets, trapped red blood cells, and fibrin (*see* FIBRINOGEN).

blood grouping a method for classifying blood types by checking which particular ANTIGENS are present on the surface of red blood cells. More than four hundred antigens can be recognized by their specificity for a particular ANTIBODY. There are many systems for the classification of blood types, but the ABO blood group and rhesus (Rh) group are two important systems that are widely known. In the ABO blood group system, there are basically two antigens, designated A and B. The A and B antigen may be present singly or together (AB), and the absence of both antigens gives rise to blood group O. There are thus four blood groups—A, B, AB or O. Naturally occurring antibodies to the ABO system develop only after the age of three months. A person of blood group A will have antibody-B present in their serum, i.e. anti-B serum. A person of blood group B has anti-A in their serum. A person of blood group O has both anti-A and anti-B, whilst a person of blood group AB has no antibodies. In blood transfusions, it is vital that blood groups are correctly matched since incompatibility will result in blood clotting, which could cause the recipient's death. For instance, if the donor is blood group A and the recipient is blood group B, then anti-B serum of the donor will react with the B-antigen present in the recipient's blood and initiate blood clotting. Of course, the A-antigen of the donor's blood will also react with the anti-A in the recipient's blood.

The rhesus blood group system can be simply explained in terms of whether an individual is Rh-positive or Rh-negative. The presence of an Rh-factor (D-antigen) on the surface of red blood cells will classify an individual as Rh-positive, whereas the absence of such a factor will classify the individual as Rh-negative.

blood plasma blood from which all the blood cells (ERYTHROCYTES, LEUCOCYTES and platelets) have been removed. The resulting solution is 90 per cent water and contains some proteins, sugar, salt, urea, hormones and vitamins.

blood serum a fraction of the liquid medium of blood, i.e. plasma, minus one of the plasma proteins, called FIBRINOGEN.

blood type *see* **blood grouping**.

BOD *see* **biological oxygen demand**.

Bohr, Niels Henrik David (1885-1962) a Danish physicist who carried out crucial research on the QUANTUM theory. In 1922 he received a Nobel prize, and in 1952 he helped establish CERN, the European organization for nuclear research, which has a laboratory in Geneva for the investigation of the physical properties of high-energy particles. He created the theory of an electron-proton atom, and his son, the physicist Aage Bohr (1922-), was part of the Nobel prize-winning team of 1975 for research on the atomic nucleus theory.

Bohr effect the discovery by the Danish physiologist Christian Bohr (1855-1911) that the oxygen-carrying capacity of blood varies within different parts of the body. He showed that the pH of the body tissue determined whether oxygen would be released from, or taken up by, the HAEMOGLOBIN in red blood cells. At low pH (acid conditions), oxygen is released from haemoglobin and enters the surrounding tissues, but at high pH (alkali conditions), oxygen from the surrounding tissue will bind strongly to the haemoglobin in the red blood cells. The Bohr effect explains why oxygen is taken up by the haemoglobin in blood circulating throughout lung tissue (high pH) but is released by blood circulating in active muscle sites (low pH).

Bohr theory a theory proposed by Niels BOHR to explain the structure of the atom. In essence, he postulated that an electron in an atom moves in circular orbits about the nucleus, in a so-called stationary state of energy. When electrons move between orbits, energy is absorbed or emitted, in the latter case as light. The theory has now been superseded by the concept of wave mechanics, which deals more adequately with complexities introduced by atoms with two or more electrons.

boiling point the temperature at which a substance

changes from the liquid state to the gaseous state. It occurs when the vapour pressure of a liquid surface equals the surrounding atmospheric pressure. The boiling point of a pure liquid is measured under the standard atmospheric pressure of 1 atmosphere (equivalent to 760mm mercury).

bond the force that holds atoms together to form a MOLECULE, a compound or a lattice. The type of bond between neighbouring atoms will be determined by the electron attraction strength of each atom and can be of three types—COVALENT, IONIC or POLAR COVALENT.

bond dissociation energy the energy required to break a specific bond that holds two atoms together in a diatomic molecule. For example, 949 kJ are needed to carry out the dissociation of a diatomic nitrogen molecule (N_2) into two separate atoms.

botulism the most dangerous type of food poisoning in the world, caused by the anaerobic bacterium called *Clostridium* botulism. This bacterium is found in an oxygen-free environment, such as underneath soil or in an airtight food can. During growth, it releases toxins that, if bacteria are living in the cells of the body, will affect the nervous system of humans. This can result in death, especially if it is the vulnerable members of a population who are affected, e.g. newborn babies and elderly people.

Boyle's law the principle that a volume of gas varies inversely with the pressure of the gas when the temperature is constant. Thus, under constant temperature, the doubling of the pressure will lead to the volume being halved (*see also* GAS LAWS). The British scientist Robert Boyle (1627-1691) not only enunciated this law but he was also one of the original founders of the oldest British scientific society, the Royal Society.

bp *abbreviation for* BASE PAIR, BOILING POINT.

breeder reactor a nuclear reactor capable of manufacturing more nuclear fuel than it consumes while maintaining a CHAIN REACTION.

Brönsted-Lowry acid any molecular or ionic substance that will donate PROTONs during a chemical reaction.

Brönsted-Lowry base any molecular or ionic substance that will accept ELECTRONs during a chemical reaction.

Brownian motion a phenomenon first discovered in 1827 by the Scottish botanist Robert Brown (1773-1858). He observed the random movement of minute particles that occurs in both gases and liquids. Brownian motion is taken as evidence to support the theory that KINETIC ENERGY is an inherent quality of all matter, as it is assumed that motion that can be seen also occurs in other substances where it cannot be seen.

budding a form of asexual reproduction in which part of the parent develops a bulge (bud) that becomes detached to form the offspring. Budding only occurs in primitive organisms, such as yeast and hydra, which have a relatively simple structure.

buffer a chemical substance capable of maintaining the pH of a solution at a fairly constant value. It does this by removing hydrogen ions (H⁺) from the solution when small amounts of an acid are added, or releasing hydrogen ions when small amounts of base are added. Most buffers are ionic compounds, usually the salt of a weak acid or base.

buffer memory a temporary area of a computer memory that can be used to store data that has to be edited or transferred to a disk.

butane the fourth member of the ALKANE family (formula $C_4 H_{10}$). It is an extremely useful compound as it is easily changed from a gas to a liquid, thus allowing it to be easily stored and used as a fuel.

byte in computing, a sequence of usually 8 or 16 BITS representing a single character or a unit of memory. The memory capacity of a computer is measured in thousands of bytes (kilobytes, KB) or millions of bytes (megabytes, MB). In the American Standard Code for Information Interchange, (ASCII), 8 bits represent a single character.

C

caffeine a PURINE that occurs in tea leaves, coffee beans and other plants. It acts as a weak stimulant to the central nervous system.

calculus a large branch of mathematics dealing with the manipulation of continuously varying quantities. The many techniques of calculus were developed by the British scientist Isaac Newton (1643-1727), and independently by the German philosopher Gottfried Wilhelm Leibniz (1646-1716), and arose from the study of natural phenomena, such as the changing speed of a falling object. Differential calculus (*see* DIFFERENTIATION) is concerned with the rate of change of a dependent variable, i.e. the maximum and minimum points, gradient, etc, of the graph of a given function. Integral calculus (*see* INTEGRATION) deals with areas and volumes, or methods of summation.

calorie a unit of quantity of heat defined as that heat required to raise the temperature of one gram of water through 1°C. The calorie has in the main been replaced as a unit by the JOULE (1 calorie = 4.186 joules).

Calvin cycle the last stage of PHOTOSYNTHESIS, named after the American biochemist Melvin Calvin (1911-) who discovered that radioactive carbon dioxide ($^{14}CO_2$) became incorporated into the carbohydrate subsequently found in the cells of plants. All the various chemical reactions involved in the Calvin cycle occur in the stroma of plant cells. These reactions are termed the DARK REACTIONS of photosynthesis, since the formation of glucose can occur without light, although it does need the products generated by the light reactions, e.g. ATP. The fixation of carbon dioxide and its conversion into carbohydrate is an energy-requiring process, as it takes six turns of the Calvin cycle to produce one molecule of glucose. The required energy is generated by the HYDROLYSIS of ATP.

cancer a disease characterized by an uncontrolled growth rate of cells, leading to the formation of tumours. If the tumour remains localized, it is termed benign and is usually harmless to the host. However, malignant tumours do not remain localized but, instead, spread throughout the body and set up a secondary growth area by a process called METASTASIS. This usually causes death in the host as the malignant tumour disrupts the essential everyday processes of the cells in the affected tissues. There is not a single cause of cancer—it is triggered by a combination of factors, including exposure to carcinogens (such as tobacco smoke), radiation, ultraviolet light, certain viruses, and the possible presence of potential cancer genes (oncogenes). Treatment of different cancers involves surgery, radiotherapy, chemotherapy and, just recently in 1987, the "magic bullet" approach, where cytotoxic drugs are labelled with one specific ANTIBODY, which will only recognize a protein present on the surface of cancer cells.

candela the SI unit of luminous intensity.

capacitance the ratio of electric charge stored within a system in relation to its electric potential. A capacitor stores electric charge (Q) according to its capacitance (C) when its potential (V) has a value of one volt. Thus a system has a capacitance of one FARAD (the unit of capacitance) when a charge (Q) of one COULOMB changes the potential by one volt (V), i.e. $C = Q/V$.

capillary action a phenomenon related to SURFACE TENSION in liquids and due to inter-molecular attraction at the boundary of a liquid, which results in liquid rising or falling in a narrow tube.

carat (1) a standard weight used for measuring stones, and now equal to 200 mg. (2) a measure of the fineness of gold where pure gold is 24 carats and 22 carat gold contains 2 parts alloy.

carbides compounds of metals with carbon which in many cases produce hard, refractory materials with metallic conductivity, e.g. TUNGSTEN carbide. Carbide tools are made of tungsten or tantalum carbide, or mixtures with nickel and cobalt and are ideal as cutting tools for hard materials at high temperatures.

carbohydrates a large group of compounds containing carbon, hydrogen and oxygen, with the general

carbolic acid

formula $C_x(H_2O)_y$. The group includes the sugars, starch, and cellulose, and form the mono-, oligo- and polysaccharides. Carbohydrates play a vital role in the metabolism of all living organisms.

carbolic acid *see* **phenol**.

carbon compounds based upon carbon (C), element 6 in the PERIODIC TABLE, these include all organic compounds and are the basis of all living matter.

carbon cycle the circulation of carbon compounds in the natural world by various metabolic processes of many organisms. The main steps of the carbon cycle are:

(1) Carbon dioxide present in air and water is taken up during photosynthesis in plants and some bacteria.

(2) The carbon accumulated in plants is later released during the decomposition of the dead plant, or of bacteria or animals that have consumed any of the plant.

(3) Carbon will also be released by the burning of fossilized plants in the form of fuels—coal, oil and gas—and during the respiration of all organisms. The concentration of carbon dioxide in the atmosphere is increasing as huge areas of tropical forests are destroyed, while the consumption of fossil fuels is rising, i.e. less PHOTOSYNTHESIS to absorb the increasing CO_2 level. This may be a factor involved in the small temperature rises throughout the world, known as the GREENHOUSE EFFECT. When there are high levels of carbon dioxide in the atmosphere, heat radiation from

the sun tends to be reflected back to earth rather than lost to space.

carbon dioxide (CO_2) a colourless gas occurring in the ATMOSPHERE due to OXIDATION of carbon and CARBON COMPOUNDS. It is the source of carbon for plants and plays a vital role in METABOLISM. It solidifies at -78.5°C and is much used as a refrigerant. It is also used in carbonated drinks and, since it does not support combustion, in fire extinguishers.

carbon monoxide (CO) a colourless gas formed during the incomplete combustion of coke and similar fuels. It also occurs in the exhaust fumes of motor engines. Carbon monoxide is poisonous when breathed in because it combines with the HAEMOGLOBIN in the BLOOD to form a stable compound. This reduces the oxygen-carrying capacity of the blood. It is a valuable industrial reagent, due to its reducing properties.

carboxyl the acid group -COOH, characteristic of the carboxylic acids where an oxygen is double-bonded and the hydroxyl (OH) is singly bonded to the carbon.

carcinogen a substance that may produce cancer (carcinoma).

cardiovascular system the organization of the heart, the arteries and veins within the human body, in which the heart and the blood vessels form a virtually closed system. The minor branches of the blood vessels supply blood to every part of the body, including the bones. The only bloodless parts of the body are dead structures such as the nails and hair. However, the nail-bed and the hair roots do require a blood supply. The only other organ without its own blood supply is the cornea (the clear window of the eye). The cornea is supplied with oxygen and nutrients by means of the tears. Other than these exceptions, every part of the body requires a constant blood supply in order to receive essential nutrients such as GLUCOSE, AMINO ACIDS, etc. Any interruption in the blood supply of the cardiovascular system causes death of that tissue.

carnivore any animal that eats the flesh of other animals. The term can also refer to the mammalian group, *Carnivora*, which includes bears, cats and dogs. Carnivorous plants are ones that trap and digest insects (using special ENZYMES) in order to obtain their nitrogen requirements.

Cartesian co-ordinates a method of representing the position of a point in space, invented by the French mathematician and philosopher, René Descartes (1596-1650). For instance, the point with Cartesian coordinates (5,2) will be found by moving 5 units along the horizontal (X-AXIS) and 2 units up the vertical axis (Y-AXIS).

catabolism the degradational processes of METABOLISM. For example, the breakdown of glucose during GLYCOLYSIS is a catabolic process.

catalyst a substance that increases the rate of a chemical reaction but can be recovered unchanged at the end of the reaction. Metal catalysts such as iron and platinum are used throughout industry to lower the ACTIVATION ENERGY of a specific process. All biological processes in animals and plants involve natural catalysts called ENZYMES.

cathode the negatively charged electrode of an electrochemical cell to which CATIONS travel and gain electrons (i.e. where REDUCTION occurs).

cathode rays a stream of electrons emitted from the negative electrode (*see above*) when electricity is passed through a VACUUM TUBE. In a piece of equipment known as the cathode-ray oscilloscope, the electrons emitted by the cathode are attracted to and hence accelerated by an anode (positive electrode). The electrons are accelerated farther down the vacuum tube between two sets of parallel plates and onto a fluorescent detecting screen. This equipment is used in televisions, in computers and in laboratories to measure the frequencies of waveforms.

cation a positively charged ion formed by an atom or group of atoms that has lost one or more electrons.

cellulose a polysaccharide occurring widely as cell walls in plants. It is found in wood, cotton and other fibrous materials and comprises chains of GLUCOSE units. It is used in the manufacture of paper, plastics and explosives.

Celsius scale *or* **centigrade scale** (C) a temperature scale with a freezing point of 0° and a boiling point of 100°, devised by the Swedish astronomer Anders Celsius (1701-44).

central processing unit *see* **CPU**.

centrifugation a technique in which a high-speed rotating machine, a centrifuge, generates centrifugal forces to separate the various components of a liquid. Different components suspended in the liquid will separate at different centrifugal speeds depending upon their size and mass.

centripetal force the product of a body's mass and the centripetal acceleration. The centripetal acceleration is the acceleration of a body moving along a curved path, directed towards the centre of that path's curvature. The centripetal force is equal and opposite to the centrifugal force.

centromere the constricted region of a CHROMOSOME, to which a pair of sister CHROMATIDs are attached. All nuclear chromosomes of EUCARYOTIC cells have a centromere, the exact position of which varies but which is genetically inheritable. The centromere can therefore be used in the identification of chromosomes, which, to be observed under the microscope, must be in cells which are undergoing either MEIOSIS or MITOSIS. The centromere is also an important factor in nuclear division as the absence of one will lead to failure of segregation; that is, one daughter cell will contain both sister chromosomes instead of one chromosome per each separate daughter cell. Thus only one cell is inheriting the genes that code for vital proteins.

CFC (*abbreviation for* chlorofluorocarbon) a chemi-

chain reaction a multi-step reaction in which the products of each step are reactants in the subsequent step. The mechanism for a chain reaction involves a slow initiation reaction followed by chain propagation until the reaction becomes inhibited and terminates.

chain rule a method used in CALCULUS to differentiate a FUNCTION of a function. By substituting another differentiable function into the original function, a COMPOSITE function is formed and differentiation can proceed using the following formula:

$$\frac{dy}{dx} = \frac{dy}{du} \times \frac{du}{dx}$$

Charles' law this states that at constant pressure, the volume of a gas varies directly with the temperature, i.e. if the temperature is doubled, then the gas volume will also double (*see also* GAS LAWS).

chelation a reaction between a metal ION and an ORGANIC molecule which produces a closed ring thus tying up the unwanted metal ion. Chelation occurs naturally in soils, removing metal ions in solution which may be potentially toxic to plants. This principle can be applied to domestic products, e.g. chelating agents are often added to shampoos to soften water by 'locking up' calcium, iron and magnesium ions (*see* HARD WATER.)

[Note: the top of the page continues from previous entry:] cal widely used in manufacturing processes, which reacts with and destroys OZONE, causing depletion of the OZONE LAYER.

chemical equation the representation of a chemical reaction using SYMBOLS for atoms and molecules.

chemical oxygen demand (COD) an indicator of water (or effluent) quality. Oxygen demand is measured chemically (*compare* BIOLOGICAL OXYGEN DEMAND) using potassium dichromate as the oxidizing agent. The OXIDATION takes two hours, providing a much quicker assessment than the BOD.

chemisorption *see* **adsorption**.

chemistry the study of the composition of substances, their effects upon one another and the changes which they undergo. The three main branches of chemistry are ORGANIC, inorganic, and physical chemistry.

chemotaxis the movement of an organism towards or away from a specific chemical. For example, some bacteria in a solution have been shown to move (using their flagella) towards an area of high glucose concentration.

chemotherapy the use of toxic chemical substances to treat diseases where the chemicals are directed against the invading organisms or the abnormal tissue.

china clay a clay composed primarily of kaolinite which is formed due to hydrothermal alteration of granite. It is extracted using high pressure water jets and is used in many industries including ceramics, paper and pharmaceuticals.

chip a tiny piece of semi-conducting material, such as silicon, printed with a microcircuit and used as part of an integrated circuit.

chi-squared test a statistical test used to determine how well data obtained from an experiment (the observed data) fits with the data expected to occur by chance. The chi-squared test is a simple method of checking that the experimental results are significant and have not just arisen from chance events.

chitin a HYDROCARBON, related to CELLULOSE but containing nitrogen, which forms the skeleton in many invertebrates, e.g. the shells of insects.

chlorofluorocarbon *see* **CFC, ozone layer**

chlorophyll the green pigments of plants present in the cell ORGANELLES called CHLOROPLASTS. Chlorophyll is an essential factor in PHOTOSYNTHESIS as it traps energy from sunlight and uses it to split water molecules into hydrogen and water.

chloroplast an ORGANELLE which is found within the cells of green plants and ALGAe and which is the site of PHOTOSYNTHESIS. The chloroplast consists of closed membrane discs called thylakoid vesicles (often stacked together to form a granum), which are surrounded by a watery matrix termed the stroma. The stroma contains enzymes necessary for the CALVIN CYCLE, and the thylakoid vesicles contain chlorophyll, an essential pigment of photosynthesis as it absorbs light energy.

cholesterol an insoluble molecule that is abundant in the plasma membrane of animal cells. In mammals, cholesterol is synthesized from saturated fatty acids in the liver and is transported by carrier molecules in the blood, called low-density LIPOPROTEIN (LDL). Cholesterol enters cells via the LDL-receptor on the plasma membrane of the cell. This mechanism of uptake helps regulate the levels of cholesterol in the blood. If a person's diet is high in cholesterol, the number of LDL-receptors on the cell membrane will decrease, which will result in a decrease in the cellular uptake of cholesterol leading to a corresponding increase in the levels of cholesterol in the blood. This excess cholesterol is deposited in the artery walls and ultimately leads to ARTERIOSCLEROSIS. Some unfortunate individuals are particularly prone to arteriosclerosis as they carry a gene that codes for a defective LDL-receptor, making the cellular uptake of cholesterol an impossible task. This inherited disorder is called familial hypercholesterolemia, which basically translates as "too much cholesterol in the blood." Cholesterol is also the main component of steroid hormones and bile salts.

chromatid one of a pair of side-by-side replica CHROMOSOMES joined by the CENTROMERE and produced during the replication of DNA within a cell.

chromatography a method for isolating the constituents of a solution by exploiting the different bonding properties of molecular substances. One complex method is affinity chromatography, which is used to separate proteins. A specific ANTIBODY is coupled to small plastic beads to form a column through which the protein solutlon is passed. The only protein that will bind to the column is the one recognized by the antibody; other proteins will pass through the column unimpeded.

chromosome a highly structured complex of DNA and PROTEINS found within the nucleus of EUCARYOTES. A nuclear chromosome is involved in the following three functions:

(1) Replication—this ensures the correct duplication of the DNA, or cells would become unviable through genetic mutation.

(2) Segregation—this ensures that the newly replicated chromosomes will separate and that each will become part of a daughter cell. It is a very important function dependent largely on the CENTROMERE.

(3) Expression—this ensures that the genes present on the chromosome are correctly transcribed to preserve the genetic information coding for specific proteins.

PROCARYOTIC cells contain a single, circular, chromosomal DNA molecule, whereas evidence indicates that EUCARYOTIC cells contain a single, linear molecule of double-helical DNA within their nucleus. However, the eucaryotic ORGANELLES, MITOCHONDRIa and CHLOROPLASTS contain a single circular molecule of DNA, which perhaps is an indication of their origin, i.e. as procaryotic cells that have evolved into eucaryotic organelles. A species will have a characteristic number of base pairs of DNA all packed into a characteristic number of chromosomes. For instance, yeast cells contain approximately 14×10^6 base pairs (bp) packed into 16 chromosomes, and the human nucleus contains approximately 3000×10^6 bp packed into 46 chromosomes. Chromosomes are visible under the microscope only during MEIOSIS or MITOSIS, when they condense to form short thick structures before separating.

cilia (*singular* **cilium**) fine thread-like structures on cell surfaces, which beat to create currents of liquid over the cell surface or to move the cell.

circuit a path that, when complete, allows electric charge to flow through it. There are only two possible types of circuit configurations—PARALLEL CIRCUITS and SERIES CIRCUITS. In series circuits, the same amount of current flows through all the components, but there is a different drop in POTENTIAL DIFFERENCE through each component. In parallel circuits, a different amount of current flows through each component, but the same drop in potential difference occurs across each component of the circuit.

citric acid cycle a complex set of biochemical reactions controlled by ENZYMES. The reactions occur within living cells, producing energy, and the cycle is instrumental in the final stages of the OXIDATION of CARBOHYDRATES and fats and is also involved in the synthesis of some AMINO ACIDS.

clone an organism, micro-organism or cell derived from one individual by asexual process. All such individuals have the same GENOTYPE. Plants propagated by cuttings are clones.

closed-chain in organic chemistry when carbon atoms are bonded together to form a ring, as typified by BENZENE. Closed-chain compounds are thus called ring or cyclic compounds (*see also* HETEROCYCLIC COMPOUNDS).

coal *see* FOSSIL FUELS.

COBOL (*acronym for* Common Business Orientated Language) a programming language geared towards a computer that is mostly used in commerce.

COD *abbreviation for* CHEMICAL OXYGEN DEMAND.

coefficient the constant numerical part of a term in an algebraic equation. For example, in the equation $5x - 6xy + y = 0$, the coefficient of x is 5, that of xy is -6, and that of y is 1. In some mathematical expressions, unknown coefficients can be represented by letters and their values calculated using a particular formula.

colligative property an aspect of a solution that depends solely on the concentration and not the nature of the dissolved particles. Colligative properties are important when determining, say, osmotic pressure (*see* OSMOSIS), vapour pressure, or the freezing point of a solution.

colloid a substance forming particles in a solution varying in size from a true solution to a coarse suspension. The particles, measuring 10^{-4} to 10^{-6} mm are charged and can be subjected to ELECTROPHORESIS.

combination the random selection of a group of objects from a given set to form a subset. The number of com-

commutative

binations for selecting a specific size of subset can be calculated using:

$$nC = \frac{(n)}{(r)} = \frac{n!}{r!\,(n-r)!}$$

where r = size of subset to be selected; n = size of original set; and ! = factorial ($3! = 3 \times 2 \times 1 = 6$)

For example, if a coach wanted to select 15 men out of a total of 20 to form an American football team, there would be

$$\frac{20!}{15! \times 5!} = 15504 \text{ (ways of selecting the team)}$$

commutative a term used to describe any mathematical operation that will give the same result independent of the order in which the elements of the operation are performed. Thus, multiplication is commutative, e.g. $2 \times 3 \times 6 = 6 \times 2 \times 3$, but subtraction is not, e.g. $6 - 3 \neq 3 - 6$.

commutator the part in a motor or ARMATURE that makes contact with the carbon brushes to carry current.

common difference the value given to the difference between successive terms in an ARITHMETIC SERIES.

common logarithm *see* **logarithm**.

complex number a number represented by the symbol z, which has the following form:
$z = a + ib$ (where a and b are real numbers and i is $\sqrt{-1}$).

composite a term used to describe a function, number or POLYNOMIAL that has factors that differ from the given function, number or polynomial. For example, $73 \times 5 = 365$ is composite and $y = x(x + 8)3$ is a composite function. PRIME NUMBERS are not composite as they are divisible only by themselves or unity.

composition of functions the mathematical operation in which a single function is formed from two given functions. The function is formed by inserting the second function into the first equation. In mathematical terminology, the composition of function "f" with function "g" is written as $f \, o \, g$. For example, the two functions, $f(x) = x + 2$, and $g(x) = x^2$ gives the following:

$$\begin{aligned} f \, o \, g &= f[g(x)] & g \, o \, f &= g[f(x)] \\ &= g + 2 & &= f^2 \\ &= x^2 + 2 & &= (x + 2)^2 \end{aligned}$$

concentration the quantity of substance (SOLUTE) dissolved in a fixed amount of SOLVENT to form a SOLUTION. Concentration is measured in moles per litre ($moll^{-1}$).

concentric a mathematical term used to describe geometrical figures that share a common centre.

condensation the process by which a substance changes from the gaseous state to the liquid state and, in so doing, loses KINETIC ENERGY.

conduction a method of heat transfer through a solid, or the flow of electrical charge through a substance. Transfer of heat energy by conduction is always from

conductor

a high temperature to a lower temperature region. A poor **conductor** of heat and electricity is called an INSULATOR, e.g. plastic, cork. All metals and carbon (GRAPHITE form) are good conductors of electric charge as they contain ELECTRONs, which are free to carry and thus transfer energy (*compare* ELECTROSTATICS).

conductor *see* **conduction**.

congruent in mathematics, a term used to describe the relationship between figures having an identical shape and size, but possibly a different orientation in space. It can also refer to a pair of numbers relative to a third number, which is indeed the difference or a multiple of the difference, between the original pair.

conservation of energy, law of *see* **energy, thermodynamics**.

contact process the inexpensive manufacture of the important industrial chemical SULPHURIC ACID (H_2SO_4). Sulphur or sulphur ores are oxidized to obtain sulphur dioxide (SO_2), which is further catalytically oxidized to sulphur trioxide (SO_3). The extremely reactive SO_3 gas is then dissolved in concentrated sulphuric acid (H_2SO_4), and the resulting pyrosulphuric acid ($H_2S_2O_7$) is diluted with a specific volume of water to give the desired concentration of sulphuric acid (H_2SO_4).

continental drift a geological concept formulated by the German geophysicist Alfred Wegener (1880-1930) that 200 million years ago the earth consisted of a large single continent, called the Pangaea, which broke apart to form the present continents. An explanation of how such huge land masses move is provided by studying the vast plates that make up the outer layer of the earth, called the crust. The crustal plates are believed to float on a partially molten region of the earth between the crust and the earth's core, the lower mantle (hence the modern term plate tectonics).

continental shelf the surface between the shoreline and the top of the continental slope where the gradient steepens at a depth of approximately 150 metres. The average width of the shelf is 70 metres.

convection a method of heat transfer through liquids and gases. It will only occur if the lower temperature area is above the high temperature area of a liquid or gas. Convection currents can be seen easily when a dye is placed at the bottom of a container of water and heat is applied to that region, causing the dye and water molecules to rise and disperse throughout the container. An everyday example of convection that can be readily detected by hand is the hot air that rises from a warm radiator.

coronary artery one of the two main vessels that carries oxygenated blood to the heart muscle. In general terms, the left coronary ARTERY divides into two branches and supplies most of the blood required by the left VENTRICLE, while the right coronary artery supplies blood to the right ventricle. The two coronary arteries originate from the AORTA, but run indepen-

cosecant

dently on to the surface of the heart.

cosecant the function of an angle in a right-angled triangle given as the reciprocal of the sine function. The cosecant of angle A is 1/sine A.

cosine a trigonometric function used to determine the value of a specific angle in a right-angled triangle by calculating the adjacent-to-hypotenuse ratio (adjacent, hypotenuse refers to the sides of the triangle).

coulomb a unit of electric charge (C): the amount of electricity carried through a given region in one second by a current of one AMPERE (A), i.e. Charge (C) = Current (A) × Time (s).

covalent bond the joining of two atoms due to the equal sharing of their electrons. In a single covalent bond, one electron pair is equally shared between two atoms (as in the hydrogen molecule, H^2). In a double bond, two electron pairs are shared, and in a triple bond, three electron pairs are shared between two atoms. Covalent bonds are usually formed between non-metallic elements in which the atoms have a strong but equal attraction for electrons, resulting in the formation of strong, stable bonds.

CPU (*abbreviation for* central processing unit) in computing, the electronic device that accepts and processes information that can be modified and stored and then output.

cracking a process much used in industry where large, complex molecules, e.g. HYDROCARBONS, are broken down into smaller molecules. Heat is usually the primary agent but pressure and CATALYSTS are also used. It is used particularly in the PETROLEUM industry.

crossing-over the reciprocal exchange of genetic material between two HOMOLOGOUS CHROMOSOMES during MEIOSIS. This process occurs at a site called the chiasma of the two chromosomes and is responsible for GENETIC RECOMBINATION.

crude oil PETROLEUM in its unrefined state.

crust *see* **continental drift**.

cryogenics the study of materials and their behaviour at very low temperatures. It is usually dependent upon the use of liquefied gases, known as **cryogens**.

cube *see* **polyhedron**.

current the flow of charge in an electrical CIRCUIT. Current (I) is measured in AMPERES (A).

cybernetics a science developed by the American mathematician Norbert Wiener (1894-1964). Cybernetics is the comparative study of organization, regulation and communication within control systems. Such studies can provide very useful information concerning the role of man and machine at work, e.g. reducing human error or devising a computer capable of making managerial decisions.

cyclic compound *see* **closed-chain**.

cytoplasm the part of the cell outside the NUCLEUS, but within the cell wall.

cytosine ($C_4H_5N_3O$) a nitrogenous base component of the NUCLEIC ACIDS, DNA and RNA, which has a pyrimidine structure and BASE PAIRS with GUANINE.

D

dark reaction the stage in PHOTOSYNTHESIS that occurs in the stroma of CHLOROPLASTS in plants and is not directly dependent on light (*see* CALVIN CYCLE).

Darwinism the theory of evolution described by the British naturalist Charles Robert Darwin (1809-1882). He first established the evolutionary theory of environmental forces acting as agents of natural selection on successive generations of organisms. The theory had five elements:

(1) Individuals have random variability, more so if they sexually reproduce.

(2) Reproductive capacity inevitably leads to competition both within (intra) and between (inter) species as populations tend to remain a set size (if there are no major external influences to upset this, e.g. mass-slaughter of animals by humans or an environmental catastrophe such as a huge oil slick).

(3) Some individuals are better adapted than others to their environment, which will help in their survival and reproductive success, i.e. fitness.

(4) Some of the characteristics that make parents successful in their ability to survive and reproduce are inherited by their offspring, thus increasing the prob-

ability of success for those offspring. These characterisitics become more widespread in the population. This is evolutionary change; the surviving species has adapted better to an environmental niche, i.e. survival of the fittest.

(5) The descendants of a single stock tend to diverge and become adapted to many different environmental NICHES. Darwin's paper, "Origin of Species" (1859), was greatly criticized by both the religious and scientific establishments of that period. He did not know of GENES as the units of inheritance, so it was understandable that he believed in the inheritance of acquired characteristics, as this would be a rational explanation for the phenomena he observed. With the knowledge that information regarding an organism's PHENOTYPE is present in their genes, then Darwin's theory can be summarized as follows—if the organisms have the capacity for genetic variability, then new species can emerge that will adapt to new environments, while the old species, which are no longer suited to the surrounding environment, will eventually die out.

DC *see* **direct current**.

decibel a unit for measuring the intensity of sound levels; it has the symbol dB.

decimal a structured system of numbers based on 10. *See also* PLACE VALUE NOTATION, and APPENDIX 4.

defect a break of discontinuity in the structural arrangement of a crystal, whether in the pattern of

ELECTRONS, ATOMS or IONS. The discontinuity may take the form of a vacancy or point defect, or be a linear break (line defect) or dislocation.

dehydration in chemistry, the removal of water that is chemically held in a molecule or compound by the action of heat, often with a CATALYST or chemical acting as a dehydrating agent. SULPHURIC ACID is an example of the latter. In medicine, dehydration is the excessive, often dangerous, loss of water from the body tissues.

delta a roughly triangular area of sediment formed at the mouth of a river. It is due to a current laden with sediment entering a body of water, resulting in a reduction of the current's velocity and thus its carrying capacity. Much of the sediment is therefore deposited on entering the lake or sea. The shape of the delta depends upon various factors including water discharge, climate, tides and sediment loads. Modern examples include the Mississippi and the Nile.

denaturation the disruption of the weak bonds that hold a PROTEIN together, usually caused by extreme heat, or addition of a strong acid or alkali. Most denatured proteins precipitate and cannot refold into their original structure. During cooking, the proteins within an egg become denatured, causing both the yolk and the white of the egg to solidify. Extremes of temperature are fatal to the majority of all animals when the protein molecules, called ENZYMES, undergo irreversible denaturation and cannot perform their essen-

detergent

tial function as CATALYSTS of the biochemical processes necessary for life.

dendrochronology the science of dating through the study of annual tree rings.

denominator the number below the line in a VULGAR FRACTION e.g. 3 in 2/3.

density the mass per unit volume of a substance. It is measured in units of kilograms per cubic metre (kgm^{-3}) and can be calculated using the following equation:

density (d) = mass (M) / volume (V)

Thus, the density of a constant volume of gas will increase if the gas is transferred to a container smaller than the initial one, but the density of the same volume of gas will decrease if the gas is transferred to a bigger container.

dependent variable in a mathematical expression, the quantity with a value determined by the other INDEPENDENT VARIABLES. For example, in the equation $y = 6x + 3$, y is the dependent variable, i.e. the value of y is dependent on the value inserted for x (the independent variable).

derivative the rate of change of a FUNCTION relative to a specific value given to the INDEPENDENT VARIABLE. The derivative of a function is obtained by DIFFERENTIATION and usually refers to the gradient of a graph. For example, the graph of the function $y = 3x^2$ has a derivative of 6x and will therefore have a gradient of 12 when $x = 2$.

detergent a soluble substance that acts as a cleaning

agent and is particularly effective in the removal of grease and oils (both HYDROCARBONS). The hydrophobic part of a detergent molecule will readily interact with hydrocarbons, whereas the hydrophilic part will readily interact with water molecules. When added to an immiscible mixture of water and, say, grease, the "dual reactive" nature of the detergent brings the water and grease together, thus allowing them to be rinsed away.

deuterium an ISOTOPE of the atom hydrogen, which constitutes 0.015 per cent of naturally occurring hydrogen. It is indicated by either the symbol D or $^{2}_{1}H$, and will react with oxygen to form heavy water, i.e. D_2O.

dextrorotatory compound any substance that, when in crystal or solution form, has an optically active property that rotates the plane of polarized light to the right (clockwise), e.g. the sugar D-glucose (D means dextrorotatory).

dialysis a method for separating small molecules from the larger ones present in a solution. Dialysis occurs in the kidneys of all vertebrates to remove the waste products of METABOLISM.

dibasic acid a substance containing two replaceable hydrogen atoms per molecule. A dibasic acid will react with a BASE to form either an acidic salt (only one hydrogen atom is replaced) or a normal salt (both hydrogen atoms are replaced). Thus the products of the neutralization of a dibasic acid depend upon the

diffusion

quantities of the acidic and basic reactants.

dicotyledon *see* **monocotyledon**.

dielectric a material, gas, liquid or solid that does not conduct an electric current—an INSULATOR. Dielectrics can be used for capacitors (*see* CAPACITANCE), terminals, and cables.

differentiation a procedure used in CALCULUS for finding the derivative of a function. There are various methods of differentiation, depending upon the simplicity or complexity of the function and its corresponding graph. One of the simplest forms of differentiation is for the common function, $f(x) = x^n$, which has the derivative $f'(x) = nx^{n-1}$. For example, the function $f(x) = 3x^3$, has the derivative $f'(x) = 9x^2$, whereas the function $y = 2x^4 + 6x^2$, has the derivative $dy = 8x^3 + 12x$. For more complex methods of differentiation *see* CHAIN RULE, PRODUCT RULE, QUOTIENT RULE.

diffraction the bending of waves round an obstacle such as the straight edge of a barrier. The diffraction of all waves—water, light, sound and electromagnetic—around a suitable object can be detected by a change in the wavefront shape. There will be no change in the wavelength or wave frequency, and consequently the speed of the wave will be constant provided the properties of the surrounding medium remain constant.

diffusion the natural process by which molecules will disperse evenly throughout a particular substance.

The molecules will always travel down their concentration gradient; that is, they will move from a region where they are highly concentrated to a region where they are lower in concentration within that substance. Diffusion occurs in gases, liquids and, depending on the size of the molecules, across the membranes of cells and ORGANELLES within cells.

digital a term signifying the use of a numerical code as opposed to the mechanical indicators on a dial. In a digital clock, illuminated numerical figures are used to display time, whereas a conventional clock has indicators that move around a dial.

dimer a molecule formed by two identical molecules bonding together. The formation of THYMINE-thymine dimers in the same strand of DNA occurs in cells damaged by exposure to ultraviolet light.

diode a device that will allow current to flow in only one direction. A diode valve has a solid surface that, when heated, will emit electrons and hence become the negatively charged plate (CATHODE). The current will travel towards a positively charged plate, the ANODE, which is connected to the positive terminal of a power source such as a battery. Overall, the arrangement of the cathode, anode and heating supply in a vacuum is called a diode valve or vacuum tube.

diploid a term used to describe a cell having two of each CHROMOSOME in its nucleus. All diploid organisms will have HOMOLOGOUS pairs of chromosomes, with each member having a similar distinctive shape.

The homologous chromosomes of a pair contain GENES that code for the same products during PROTEIN synthesis, although the information could result in a different form of protein and thus a characteristic (*see* ALLELE). Humans are diploid, with each cell of the body (except GAMETEs) having 46 chromosomes; that is, 22 pairs of homologous AUTOSOMES and a pair of sex chromosomes.

dipole one of two equal and opposite charges (electric dipole) or magnetic poles (magnetic dipole), separated by a short distance. In an ATOM, a transient electric dipole is generated by the random distribution of its continuously moving electrons, and in a MOLECULE a dipole is the result of unequal sharing of electrons within a bond. If two atoms of a molecule differ in ELECTRONEGATIVITY, one atom will have a greater attraction for the electrons of the bond than the other. This will cause one end of the molecule to be slightly positive and the other end to be slightly negative. For example, water molecules (H_2O) are dipoles, as the greater electronegativity of the oxygen atom (3.5) will have a stronger attraction for the electrons of the bonds than does the electronegativity of the hydrogen atoms (2.1). Thus, the bonding electrons spend more time around the oxygen atom, giving the oxygen end a slightly negative charge and the hydrogen end of the H_2O molecule a slightly positive charge.

dipole (*see* POLAR COVALENT BOND).

direct current (DC) an electric current flowing in one

discriminant

direction only.

discriminant *see* **quadratic equation**.

displacement the calculation determining the change of position of an object by relating its final position to its initial position. Displacement is always independent of the path taken and is measured in metres (m).

dissociation the process by which a compound breaks up into smaller MOLECULES, or IONS and ATOMS.

distance in physics, the measurement of how far an object has travelled along a specific path. In mathematics, distance is the length of the line needed to join particular points and is similar to displacement. The units of distance are metres.

distillation the separation of a liquid solution into its various components by initially heating the liquid into vapour and then cooling the vapour so that it condenses and can be collected. The different components of a solution can be collected at different intervals as each component will have a unique BOILING POINT.

distribution law the ratio of the concentrations of the SOLUTEs dissolved in two immiscible liquids, which are in equilibrium with each other. The value is constant when the temperature is also constant. The temperature must be constant as heat increases the solubilities of liquids in liquids.

dizygote a term used to describe twins who have developed from two separate ova and therefore have different genetic characters (*see also* FRATERNAL TWINS).

DNA (*abbreviation for* **deoxyribonucleic acid**) a NUCLEIC ACID and the main constituent of chromosomes. Made up of a double helix of two long chains of linked NUCLEOTIDES, BASE PAIRed between ADENINE and THYMINE or CYTOSINE and GUANINE, it transmits genetic information in the form of GENES from parents to offspring.

DNA polymerase an enzyme that helps in the replication of new DNA strands by using the uncoiled double-helix of DNA as a template on which to add free NUCLEOTIDES. Three DNA polymerases have been identified, and all of these enzymes help catalyse the step-by-step addition of a DNA nucleotide to the end of a growing DNA chain:

DNA Polymerase I—gap filling and repair enzyme
DNA Polymerase II—function unknown at present
DNA Polymerase III—responsible for the majority of DNA synthesis

dominance a genetic concept that a certain ALLELE of a gene will mask the expression of another allele known as the recessive. The PHENOTYPE of the individual is a result of the expression of the dominant allele with the recessive allele (*see* HETEROZYGOTE, HOMOZYGOTE) remaining undetected as it will have no effect on the phenotype.

doppler effect a change in the observed frequency of a wave as a result of the relative motion between the wave source and the detector. The doppler effect is responsible for an ambulance siren having a higher

double bond the linking of two atoms through two COVALENT BONDS in a compound, i.e. the sharing of two pairs of electrons.

dry ice the solid form of carbon dioxide (CO_2), which is used as a refrigerant. It sublimes (*see* SUBLIMATION) at -78°C without forming a liquid, hence the term "dry" ice.

ductility the property of metals of alloys that enables them to be drawn out into a wire and to retain strength when their shape is changed.

duodenum the first part of the small intestine of vertebrates that connects the stomach to the ileum and into which secretions from the gall bladder and pancreas are emptied.

dynamics the branch of physics that deals with the motion of objects under the action of the forces responsible for changes in the motion of those objects.

dynamo a machine that generates electric currents using a rotating coil as a conductor and powerful magnets to create a MAGNETIC FIELD.

E

ECG (*abbreviation for* **electrocardiograph**) equipment used to record the current and voltage associated with contractions of the heart.

eclipse the total or partial disappearance of one astronomical body by passing into the shadow of another. A solar eclipse occurs when the new moon passes between the earth and the sun. A lunar eclipse occurs when the moon moves into the shadow of the earth, i.e. the earth is situated between the sun and the moon.

ecology the study of the relationship between plants and animals and their environment. Ecology is also known as bionomics and is concerned with, for example, predator-prey relationships, population dynamics and competition between species.

ecotype a population that is genetically adapted to the local conditions of its particular habitat.

ecosystem an ecological community that includes all organisms which occur naturally within a specific area.

Einstein, Albert (1879-1955) a German-born American physicist who formulated the RELATIVITY theories and carried out important investigations in THERMODYNAMICS and radiation physics. In 1921, he received the Nobel Prize in physics. He became a professor of mathematics at Princeton University, New Jersey, in 1933 after fleeing Nazi Germany.

elasticity a property of any material that will stretch when forces are applied to it and recover when the forces are relaxed. To stretch a spring or any other elastic material, equal but opposite (direction) forces must be applied to two areas of the material. All materials are elastic to some extent and will obey HOOKE'S LAW if the forces applied do not cause permanent deformation.

electric current a flow of electric charge through a CONDUCTOR. The charge may be carried by means of electrons, ions or holes. Hole is the term for the absence of an electron in the valence structure (*see* VALENCY) of a body. The movement of an electron in to the hole creates new holes and therefore "conduction by holes" (*see also* CURRENT).

electric field the invisible force that always surrounds any charged particle. When one charged particle is within proximity of the electric field of another, each will feel and exert a force. If the charged particles are opposite (unlike), then they will attract since opposite charges attract. However, two particles with the same charge, i.e. both positive or both negative, will repel each other. An electric field is diagrammatically represented by lines along which a free positive charge would theoretically move, and so the arrows will always point towards a negative charge and away from a positive charge. The strength of an electric field is represented by the density of the drawn lines, but can be measured in either volts per metre (Vm^{-1}) or newtons per coulomb (NC^{-1}).

electrocardiograph *see* **ECG**.

electrolysis chemical decomposition achieved by passing an electric CURRENT through a substance in solution or molten form. IONS are created, which move to electrodes of opposite charge where they are liberated or undergo reaction.

electrolyte a compound that dissolves in water to produce a SOLUTION that can conduct an electrical charge. The electrical conductivity of the solution is a result of the compound undergoing ionization to form mobile IONS. If the substance is completely ionized, it is termed a strong electrolyte, e.g. any strong acid such as sulphuric acid, but if it is only partially ionized it is termed a weak electrolyte, e.g. any weak acid such as ethanoic (acetic) acid.

electromagnetic induction the production of an ELECTRIC CURRENT when a conductor is moved through a MAGNETIC FIELD. A current is induced to flow only while there is a changing magnetic field due to the movement of the conductor or magnet. The direction of the induced current flow depends on the pole orientation of the magnet, i.e. north-south or south-north, and whether the magnet is moving towards or away from the conductor. The magnitude of an induced current is dependent on the strength of the magnetic field,

electromagnetic waves

the cross-sectional area of the conductor, and both the speed and direction of the relative motion between the conductor and the magnetic field.

electromagnetic waves the effects of oscillating electric and magnetic fields that are capable of travelling across space, i.e. they do not require a medium through which to be transmitted. The spectrum of electromagnetic waves is divided into the following categories:

	wavelength (m)	frequency (Hz)
gamma rays	10^{-10}-10^{-12}	10^{21}- ......
X-rays	10^{-12}-10^{-9}	10^{21}-10^{17}
ultraviolet radiation	10^{-10}-10^{-7}	10^{18}-10^{15}
visible light	10^{-7}-10^{-6}	10^{15}-10^{14}
infrared radiation	10^{-6}-10^{-2}	10^{14}-10^{11}
microwaves	10^{-3}-10	10^{11}-10^{7}
radiowaves	10 - 10^{6}	10^{7}-10^{2}

All electromagnetic waves travel through free space at a speed of approximately 3×10^8 ms^{-1}, known as the SPEED OF LIGHT. The only electromagnetic waves that are readily detected by the eye are visible light waves. These consist of various wavelengths, which correspond to the colours red, orange, yellow, green, blue, indigo and violet. The colour red has the longest wavelength, lowest frequency, and the colour violet has the shortest wavelength, highest frequency. An object that appears to be red in colour, for example, has absorbed all the light waves from the blue end of the spectrum while reflecting the ones from the red end.

electron an indivisible particle that is negatively charged and free to orbit the positively charged NUCLEUS of every atom. In the traditional model, electrons move around in concentric shells. However, the latest concept, based on quantum mechanics, regards the electron moving around the nucleus in clouds that can assume various shapes, such as a dumb-bell (two electrons moving) or clover leaf (four moving electrons). The shape and density of the outermost electronic shell will help determine what reactions are possible between particular atoms and molecules, e.g. whether an atom will easily gain or lose electrons to form an ION.

electron affinity the energy change associated with the capture of an electron by a single, gaseous atom (*compare* ELECTRONEGATIVITY). The process is EXOTHERMIC when a single electron is captured by a neutral atom (first electron affinity), but any further electron captures (second or higher electron affinities) can be ENDOTHERMIC processes.

electronegativity (*plural* **electronegativities**) a measure of the power of an atom within a MOLECULE to attract electrons. Every element in the PERIODIC TABLE is given an electronegativity rating based on an arbitrary scale in which fluorine is given the highest rating of 4, as it has the most electronegative atoms. Electronegativity differences between the different atoms within a molecule can be used to estimate the

electron pair

nature of the bonds formed between those atoms, i.e. whether it is a COVALENT, IONIC or POLAR COVALENT BOND.

electron pair two electrons from the outer shells of two atoms which are shared by the adjacent nuclei to form a bond.

electrophile a molecule that will readily accept electrons from another molecule. An electrophile usually has an electron-poor site as its functional group, as in the ion, NO_2^+, or the acid, HBr, and will thus attack the high electron density region of other molecules, such as the double bond in ALKENES.

electrophoresis a method for separating the molecules within a solution using an electric field. The molecules will move at a speed determined by the ratio of their charge to their mass. All electrophoretic separations involve placing the mixture to be separated onto a porous, supporting medium, such as filter paper, or a plant-derived gel called agarose that has been soaked in a suitable BUFFER. An electric field is applied, causing the molecules, now dissolved in the conducting buffer solution, to move at a rate according to their charge-to-mass ratio. The various molecules can be identified by comparing their final position on the supporting medium with the position of known standards. Gel electrophoresis uses a synthetic polymer, such as polyacrylamide, as the supporting medium, and can be used to separate the different lengths of DNA chains found within the cellular nucleus of an individual. This makes gel electrophoresis an essential technique of "genetic fingerprinting," as the separated DNA strands will contain genes unique to a particular person, thus helping in identification.

electrostatics the part of the science of electricity dealing with the phenomena associated with electrical charges at rest. Any material containing an electric charge that is unable to move from atom to atom is called an INSULATOR (as opposed to a CONDUCTOR, which allows elecric charge to flow throughout it). It is possible to negatively charge an insulator such as polythene by transferring electrons to it using a woollen cloth. A similar effect is sometimes produced when a plastic comb is used to comb dry hair.

element a pure substance that is incapable of separating into a simpler, different substance when subjected to ordinary chemical reactions. There are 105 elements known to us, but only 93 of these occur naturally, and the others have been created in laboratories. The elements are classified into the PERIODIC TABLE, according to the number of protons in their nucleus, i.e. their atomic number. Each element of the periodic table consists of atoms with a unique number of protons in their nuclei.

elementary particle (*or* **fundamental particle**) a particle which is the basic building block for all matter. To explain nuclear interactions fully, the NEUTRON, PROTON and ELECTRON have been supplemented by new particles, some more elementary than others. Two types of particles are thought to exist, namely leptons and hadrons, which are differentiated by the way they interact with other particles. Leptons include the electron and the neutrino, the latter possessing spin but no charge or mass. It is associated with BETA DECAY. Hadrons include protons and neutrons, which are not truly elementary, and it is postulated that these are composed of elementary particles called quarks. Quarks have become part of an elaborate theory of hadron structure encompassing properties termed "flavour" and "colour charge." Although the theory is generally accepted by physicists, quarks have not been confirmed experimentally.

embryo the developmental stage of animals and plants that immediately follows fertilization of the egg cell (ovum) until the young hatches or is born. In animals, the embryo either exists in an egg outside the body of the mother or, as in mammals, is fed and protected within the uterus of the mother.

empirical formula a chemical formula of a compound that shows the simplest ratio of atoms present in the compound. For example, the molecule BUTANE (fourth member of the alkane family) has the empirical formula C_2H_5 although its true molecular formula is C_4H_{10}.

emulsion in chemistry, a COLLOIDal solution of one liquid in another.

endocrine system the network of glands that secrete signalling substances (HORMONES) directly into the bloodstream. This enables the secreted hormone to travel to, and thus affect, distant target cells, as opposed to just affecting cells that surround the endocrine gland. The pituitary gland (at the back of the head, base of the brain), the thyroid gland (in the neck) and the adrenal glands (above both kidneys) are three of the major endocrine glands within the human body. Each secretes different hormones, which will subsequently affect different parts of the body. For example, the pituitary gland secretes growth hormone, the thyroid gland secretes thyroxine, and the adrenal glands secrete the stress hormone called glucocorticoids, all of which control various functions in the body.

endoplasmic reticulum a network of membrane-bound tubules and flattened sacs connected to the membrane of the nucleus found within all EUCARYOTIc cells. If the endoplasmic reticulum (ER) has RIBOSOMES attached to the outside membrane, then it is called "rough ER," but in the absence of ribosomes it is called "smooth ER," or GOLGI APPARATUS. Both the rough ER and golgi apparatus have functions important in the biosynthesis of other organelles and in the synthesis, modification and sorting of proteins.

endothermic reaction a chemical reaction in which heat energy is absorbed. The required heat energy is supplied by the environment surrounding the reaction, and the products of an endothermic reaction will have stronger bond energies than the original reactants.

energy the capacity to do WORK. There are various forms of energy, including light, heat, sound, mechanical, electrical, kinetic and potential, but all are expressed in the same unit of measurement, called the JOULE (J). Energy has the capacity to change from one form to another (**energy transfer**), but the original input of energy tends to be greater than the final output during energy transfers. As the law of conservation of energy states that it is impossible to make or destroy energy, the difference in the input/output energy levels is a result of the conversion of some of the input energy into an unwanted form, e.g. heat instead of mechanical energy. The energy content of a system or object can be regarded as the "work done" by it and can be calculated using the following equation:

Work Done (w) = Force (F) x Distance Moved (S)

enthalpy the quantity of heat energy (THERMODYNAMICS) possessed by a substance. Enthalpy (H) has units of joules per mole ($Jmol^{-1}$) and is defined by the equation:

$H = E + PV$ where E = internal energy of a system
P = pressure
V = volume

The enthalpy change (∂H) during a reaction is referred to as ∂H negative when the reaction is EXOTHERMIC (heat evolved) and ∂H positive when the reaction is ENDOTHERMIC (heat absorbed).

entropy a measure of the randomness (disorder) of a

epilepsy

system. It is a natural tendency of the whole universe that allows all energy to be distributed. Entropy (S) has units of joules per kelvin per mole ($JK^{-1} mol^{-1}$) and can be related to ENTHALPY (H) using GIBB'S LAW (G = H - TS). The greater the disorder, the higher the value for entropy, but at absolute zero entropy is also zero.

enzyme any PROTEIN molecule that acts as a natural CATALYST and is found in the bodies of all bacteria, plants and animals. Enzymes are essential for life as they allow the complex chemical reactions of biochemical processes to occur at the relatively low temperature of the body. Enzymes are highly specific in that they will only act on certain SUBSTRATES at a specific pH and temperature. For example, the digestive enzymes called amylase, LIPASE and trypsin will only work in alkaline conditions (pH > 7), whereas the digestive enzyme, pepsin will only work in acidic conditions (pH < 7).

epilepsy a seizure disorder caused by lesions in the brain. The symptoms are in the form of attacks, known as fits, which can include a feeling of numbness, muscular convulsions, inability to speak, etc. Epilepsy can be controlled by certain drugs, but in bygone days, surgery was performed on patients who suffered frequent and extreme attacks in an attempt to control these. The operation disconnected the left and right hemispheres of the brain by cutting the communication system between them, a fibrous network called the *corpus callosum*.

equilibrium

equilibrium a condition in which the proportion of reactants and products is constant as the rate of the forward reaction equals the rate of the reverse reaction. An equilibrium constant (K_c) can be calculated for reactions by dividing the concentrations of products from the forward reaction by the concentration of the reactants from the reverse reaction. Both the concentration of the products and reactants are raised to the power corresponding to their coefficient in the chemical equation for the reaction, i.e. for the reaction:

$$aA + bB \rightleftharpoons xX + yY$$

the equilibrium constant (K_c) is calculated as follows:

$$K_c = \frac{[X]^x [Y]^y}{[A]^a [B]^b}$$

K is expressed in moles per litre (mol^{-1}) and is useful for indicating whether a particular reaction is irreversible (large K_c) or the reactants are unreactive (small K_c). Equilibrium is affected by changes in pressure, temperature and concentration, but is not affected by the addition of a CATALYST (this affects only the reaction rate).

ER *see* **endoplasmic reticulum**.

erythrocyte the red blood cell of vertebrates that is made in the bone marrow. It differs from other cells in the human body in that just before it is released into the bloodstream, it sheds its nucleus. Erythrocytes contain HAEMOGLOBIN, the protein molecule essential for transportation of oxygen from the lungs to all tissues in the body. Within the erythrocyte membrane there is a complex molecule called the Band 3 protein, which is essential for the transport of carbon dioxide (CO_2) from all body tissue to the lungs as it allows CO_2, in the form of the bicarbonate ion (HCO_3^-), to leave the red blood cell in exchange for a chloride ion (Cl). A person deficient in the number of erythrocytes circulating in their blood is said to be anaemic. Anaemia can be a result of a defect in the structure of the erythrocyte membrane or lack of iron to form the haem group of the haemoglobin molecule.

ester an organic compound formed from acids by replacing the hydrogen with an alkyl radical, e.g. CH_3COOH (ethanoic acid) to $CH_3COOC_2H_5$. Many esters have a fruity smell and are used for flavourings. Esters are common in nature, as animal fats and vegetable oils are formed from mixtures of esters.

ethane the second member of the homologous series of ALKANES. It is an insoluble colourless gas with chemical formula CH_3CH_3. Ethane has no reactive functional group, and is, therefore, a stable molecule that will not react with ACIDS, BASES, ELECTROPHILES or NUCLEOPHILES, but, however, it will undergo a slow SUBSTITUTION reaction.

ethanol a derivative of ETHANE, which has a functional hydroxyl group (OH) in place of one hydrogen atom, i.e. CH_3CH_2OH. Ethanol is obtained either by fermentation of carbohydrates to form alcoholic bever-

ethene

ages, or commercially prepared from ETHENE by adding water and sulphuric acid.

ethene the first member of the ALKENE family, which is an insoluble gas at room temperature. Ethene (C_2H_4) is an important precursor in the industrial manufacture of the plastic POLYMER called polythene, i.e. polyethene.

ethyne the first member of the ALKYNE family, it has the chemical formula C_2H_2. It is also known as acetylene and is a highly flammable gas which, when burned with oxygen, will produce the high temperature flame (> 2500°C) characteristic of the oxyacetylene torch needed to cut and weld metals.

eucaryote any member of a class of living organisms (except viruses) that has a membrane-bound nucleus within its cells. All eucaryotic cells contain ORGANELLES, which are also bound by closed, phospholipid membranes, e.g. chloroplast, endoplasmic reticulum, mitochondrion etc. All plants and animals are eucaryotes, but bacteria and cyanobacteria are PROCARYOTES.

eugenics the study of how the inherited characteristics of a human population can be improved by genetics, i.e. controlled breeding.

evaporation the process by which a substance changes from a liquid to a vapour. Evaporation occurs when a liquid is heated and some molecules near the surface of it eventually have enough KINETIC ENERGY to overcome the attractive forces of the remaining molecules and escape into the surrounding atmosphere. During evaporation from an open container, the temperature of the liquid falls until heat from the surroundings flows in to replace this heat loss. This explains why swimmers can feel chilled when they emerge from the water; heat energy from the skin is converted to kinetic energy, allowing some water molecules to evaporate.

evolution the process by which an organism changes and thus attains characteristics distinct from existent relatives. Any species of organism will only evolve if:

(a) There has been genetic mutation allowing variation in the genetic information the parent passes on to its descendants.

(b) An individual proves to be more suitable to a particular environment than its relatives, allowing it to survive and propagate whereas its relatives will become extinct, i.e. NATURAL SELECTION (*see also* DARWINISM).

exothermic reaction a chemical reaction in which heat energy is released to the surrounding environment. The products of an exothermic reaction will have weaker bond energies and therefore be more stable than the bonds within the molecules of the original reactants.

exponent a symbol, usually numerical, that appears as a superscript to the right of a mathematical expression and indicates the power to which the expression has to be raised. For example, the expressions a^5 and 3^7 have the exponents 5 and 7 respectively.

exponential function

exponential function a mathematical function in which the constant quantity of the expression is raised to the power of a variable quantity, i.e. the exponent. For example, the exponential function, $f(x) = a^{2x}$, has $2x$ as its variable exponent. However, the term exponential function mainly refers to the function $f(x) = e^x$, in which e has a value equal to 2.7182818. The exponential function e^x is the inverse of the natural LOGARITHMIC FUNCTION, $\ln(x)$.

exponential series the sum of the infinite sequence of exponential terms for either the real exponential function (e^x) or the complex exponential function (e^z). It is defined by the following:

$$e^x = \sum_{r=0}^{\infty} \frac{x^r}{r!} = 1 + \frac{x}{1!} + \frac{x^2}{2!} + \frac{x^3}{3!} + \cdots + \frac{x^r}{r!} + \cdots$$

In the above, X represents a real function and therefore the function e^x will tend to 0 as X tends to negative infinity ($-\infty$), that is, as $X \longrightarrow -\infty$, $e^x \longrightarrow 0$. The term complex function denotes the presence of a complex number ($z = a + ib$) within the function.

extrapolate to predict the unknown value of a measurement or function using known values. On a graph, extrapolation involves extending the curve of the function beyond the set of known values for the x and y co-ordinates.

F

factor one of two or more quantities that produce a given quantity when multiplied together. For example, the factors of the number 8 are 1,2,4,8.

factorial the product of a series of consecutive INTEGERs from 1 to n inclusive, where n is a whole number. Thus, factorial 5, written as $5! = 1 \times 2 \times 3 \times 4 \times 5 = 120$. Factorials of much larger numbers are not usually defined.

Fahrenheit (F) a temperature scale, devised by the German physicist Gabriel Fahrenheit (1686-1736), that set the freezing point of water at 32° and the boiling point at 212°. Fahrenheit temperatures can be converted to CELSIUS by the equation $F = 1.8C + 32$.

farad the unit of CAPACITANCE, usually denoted by the symbol F. For example, a capacitance (C) such that a charge of one COULOMB raises the potential to one volt, is said to be one farad, i.e. 1 farad = 1 coulomb per volt ($1F = 1\ CV^{-1}$). As the farad itself is usually too large a quantity for most applications, the practical unit is the microfarad (μF).

faraday the quantity of charge carried by one mole of electrons ($\approx$ Avogadro's constant x charge on an electron), with the value 9.6487×10^4 coulombs. It was named after Michael Faraday (1791-1867), a British scientist whose contributions to physics and chemistry include ELECTROMAGNETIC INDUCTION, electrolysis and MAGNETIC FIELDS.

fatigue of metals the structural failure of metals due to the repeated application of STRESS, which results in a change to the crystalline nature of the metal.

fats a group of naturally existing lipids that occur widely in plants and animals and serve as long-term energy stores. A fat consists of a GLYCEROL molecule and three FATTY ACID molecules, collectively known as a triglyceride, which is formed during a condensation reaction (water is released). Fats are important as energy-storing molecules since they have twice the calorific value of carbohydrates. In addition, they insulate the body against heat loss and provide it with cushioning, which helps protect against damage. In mammals, a layer of fat is deposited beneath the skin (subcutaneous fat) and deep within the tissues (adipose tissue) and is solid at body temperature due to the high degree of saturation. In plants and fish, the fatty acids are generally less saturated and as such tend to have a liquid-like consistency, i.e. oils, at room temperature.

fatty acids a class of organic compounds containing a long hydrophobic (water insoluble) hydrocarbon chain and a terminal carboxylic acid group (COOH) which is extremely hydrophilic (water soluble). The chain length ranges from one carbon atom (HCOOH;

methanoic acid), to nearly thirty carbon atoms, and the chains may be SATURATED or UNSATURATED. As chain length increases, melting points are raised and water solubility decreases. However, both unsaturation and chain branching tend to lower melting points. Fatty acids have three major physiological roles:

(1) They are building blocks of phospholipids (lipids containing phosphate) and glycolipids (lipids containing carbohydrate). These molecules are important components of biological membranes, creating a lipid bilayer which is the structural basis of all cell membranes.

(2) Fatty acid derivatives serve as hormones and intracellular messengers.

(3) Fatty acids serve as fuel molecules. They are stored in the CYTOPLASM of many cells in the form of triglycerides (three fatty acid molecules joined to a glycerol molecule) and are degraded, as required, in various energy-yielding reactions.

feedback mechanism a control mechanism that uses the products of a process to regulate that process by activating or repressing it. Almost all homeostatic mechanisms (*see* HOMEOSTASIS) in animals operate by negative feedback, whereby a variation from the normal triggers a response that tends to oppose it. For example, it operates during hormonal release to maintain steady blood sugar levels. Positive feedback is found less often as a biological control mechanism. Here, a variation from the normal causes that varia-

fermentation

tion to be amplified, and this is usually a sign that the normal control mechanisms have broken down.

fermentation a form of ANAEROBIC RESPIRATION, which converts organic substances into simpler molecules, generating energy in the process. Fermentation, carried out by certain organisms such as bacteria and YEASTS, is the conversion of sugars to alcohol in the process known as alcoholic fermentation. Lactic acid fermentation occurs in the muscles of higher animals when the oxygen requirement exceeds the supply and sugar is converted into lactic acid. In industry, fermentation is important in baking and in beer and wine production, and these use large quantities of yeast.

Fermi, Enrico (1901-1954) an Italian-born American physicist who was awarded the Nobel Prize in 1934 for his discovery that stable elements would become unstable when bombarded with NEUTRONS as they have become radioactive. His later research contributed to the harnessing of atomic energy and to the construction of the atomic bomb.

fertilization the fusion of male and female GAMETES to produce a single cell, which sets in motion a chain of events that gives rise to a new individual. In animals, where the gametes unite outside the parents' bodies, it is termed external fertilization (as in most fish). Where the male gametes are deposited within the body of the female by the male, it is termed internal fertilization, as is the case with mammals. In flowering plants, after pollen has been transferred from the male

fission track dating

to the female part of the flower, a pollen tube develops, which transfers two male nuclei to the ovule of the female. Double fertilization occurs, producing a DIPLOID ZYGOTE and a triploid endosperm, which act as a food supply for the developing embryo.

fertilizer a substance added to soil to replace nutrients removed by plants, thus contributing towards their health and vitality. Fertilizers may be natural (e.g. manure) or synthetic, the latter containing nitrogen, phosphorus and potassium as the main constituents.

fibrinogen a blood PROTEIN, which causes BLOOD CLOTTING due to action by the ENZYME thrombin. The end product is fibrin.

field *see* **electric field, magnetic field.**

fission the spontaneous or induced splitting of a heavy nucleus (such as uranium) into two fragments during a nuclear reaction, which subsequently releases vast quantities of energy. Nuclear fission is induced by irradiating nuclear fuels like uranium with NEUTRONS in a device called a nuclear reactor, and this process is accompanied by the emission of several neutrons. These neutrons in turn cause fission of another nucleus, which, under suitable conditions, can result in a CHAIN REACTION. The energy released is in the form of heat, and it can be harnessed and used to produce electricity by making steam, which is used to drive turbines.

fission track dating a technique used to date some minerals and natural glasses. The FISSION of ^{238}U creates charged particles, which leave a trail through the solid glass. The tracks can be studied using a microscope, and their number relates to the age of the specimen and its uranium concentration, providing certain physical criteria have been met.

flagella (*singular* **flagellum**) long thread-like extensions from the surface of a cell, similar to cilia. Protozoa (single-celled animals) use flagella for locomotion.

flame test a simple test for the detection of metals and useful for distinguishing between different metals. A small quantity of the unknown sample is placed on a platinum wire and held in a flame, and the resultant colour is characteristic of the particular metal. For example, when sodium compounds are held into a flame, the flame burns with a bright yellow colour. Potassium gives a violet flame, and lithium and strontium give a red flame. Although lithium and strontium appear similar, the light from each can be resolved (separated) into different colours by using a prism, and this resolution easily distinguishes the two elements.

Fleming, Sir Alexander (1881-1955) a Scottish bacteriologist who discovered the antibiotic penicillin in 1928, for which he was awarded the Nobel Prize in 1945.

fluid any substance that flows easily and alters its shape in response to outside forces. All gases and liquids are fluids. In liquids, the particles move freely but are restricted to the one mass, which occupies almost the same volume. In gases, however, the particles tend to expand to the limits of their containing space and thus do not keep the same volume.

fluid-mosaic the name given to the model that describes the structure of the cell membrane in organisms, proposed by Singer and Nicholson in the 1970s. Using electron microscopy, they confirmed that the lipid component is organized in a regular bimolecular structure with protein molecules arranged irregularly along the lipid layers. Both lipid components can move laterally in the membrane.

fluorescence the property of certain substances to re-emit absorbed radiation as visible light. This occurs when molecules in the ground state are excited by incident light of a particular WAVELENGTH, thus raising their energy level. When the energy level falls, the radiation is emitted at a different, usually greater, wavelength.

flux a substance added to a solid to assist in its fusion, e.g. cryolite (naturally occurring sodium aluminium fluoride) is added to bauxite from which ALUMINIUM is extracted.

foetus the developing young in the mammalian uterus, from the post-embryonic period until birth. In humans, this period is from about 7-8 weeks after FERTILIZATION.

folic acid a compound which forms part of the VITAMIN B complex. It is involved in the BIOSYNTHESIS of

force

some ANIMO ACIDS and is used in the treatment of anaemia.

force the push or pull exerted on a body, which may alter the state of motion by causing the velocity of the body to increase or decrease. An object will continue to move at a constant speed in a straight line unless another force acts upon it. The unit of force is the newton, given by F = ma where m is the MASS of the body and a is its ACCELERATION.

formula (*plural* **formulae**) a law or fact used in science and mathematics, denoted by certain symbols or figures. In mathematics and physics, it is expressed in algebraic or symbolic form. In chemistry, there are three principal types of formulae—EMPIRICAL FORMULA, MOLECULAR FORMULA, and STRUCTURAL FORMULA.

fossil the remains of once-living plants and animals, or evidence of their existence, preserved in the strata of the earth's crust. Palaeontology is the name given to the study of fossils and has proved useful in the study of evolutionary relationships between organisms, and in the dating of geological strata.

fossil fuels these are NATURAL GAS, PETROLEUM (oil) and coal, which are the major fuel sources today. They are formed from the bodies of aquatic organisms that were buried and compressed on the bottoms of seas and swamps millions of years ago. Over time, bacterial decay and pressure converted this organic matter into fuel.

fossil fuels

Hard coal, which is estimated to contain over 80 per cent carbon, is the oldest variety and was laid down up to 250 million years ago. Another, younger variety (bituminous coal) is estimated to contain between 45 per cent and 65 per cent carbon. The fuel values of coal are rated according to the energy liberated on combustion. Coal deposits occur in all the world's major continents, and some of the leading producer countries are the United States, China, Russia, Poland and the United Kingdom.

Natural gas consists of a mixture of HYDROCARBONS, including METHANE (85 per cent), ETHANE (about 10 per cent) and PROPANE (about 3 per cent). However, other compounds and elements may also be present, such as carbon dioxide, hydrogen sulphide, nitrogen and oxygen. Very often, natural gas is found in association with petroleum deposits. Natural gas occurs on every continent, the major reserves being found in Russia, the United States, Algeria, Canada and in counties of the Middle East.

Petroleum is an oil consisting of a mixture of HYDROCARBONS and some other elements (e.g. sulphur and nitrogen). It is called crude oil before it is refined. This is done by a process called FRACTIONAL DISTILLATION, which produces four major fractions:
(1) Refinery gas, which is used both as a fuel and for making other chemicals.
(2) Gasoline, which is used for motor fuels and for making chemicals.

fraction

(3) Kerosine (paraffin oil), which is used for jet aircraft, for domestic heating and can be further refined to produce motor fuels.

(4) Diesel oil (gas oil), which is used to fuel diesel engines.

The known residues of petroleum of commercial importance are found in Saudi Arabia, Russia, China, Kuwait, Iran, Iraq, Mexico, the United States, and a few other countries.

Together, the fossil fuels account for nearly 90 per cent of the energy consumed in the United States. As coal supplies are present in abundance compared with natural gas or petroleum, much research has gone into developing commercial methods for the production of liquid and gaseous fuels from coal.

fraction a quantity that is only part of a whole unit. It is written as x/y where x is an integer and y is a natural number.

fractional distillation (*also called* **fractionation**) the process used for separating a mixture of liquids into component parts (fractions) by distillation. The liquid to be separated is placed in a flask or distillation vessel to which a long vertical column (fractionating column) is attached. The liquid is boiled, causing it to vaporize, and as the vapour rises up the column it condenses and runs back into the vessel. The vapour in the vessel continues to rise, and as it does so it passes over the descending liquid. This eventually creates a steady temperature gradient, with temperature decreasing towards the top of the column. The components of the mixture that vaporize easily (low BOILING POINTS) are said to be more volatile and are found towards the top of the column, where the temperature is lowest, while less volatile components are found towards the bottom of the column. At various points on the column, the different fractions can be drawn off and collected. Those components with appreciably different boiling points will be separated into the different fractions. Petroleum contains a mixture of hydrocarbons, and fractional distillation is used to separate the components into fractions such as gasoline and kerosine. It was fractional distillation of liquid air that led to the discovery of three of the NOBLE GASES (neon, krypton and xenon), and today the process is used to obtain large quantities of molecular oxygen required for commercial purposes. Air is first compressed and cooled (which freezes out carbon dioxide and water) and then fractioned in a liquid air machine. Molecular oxygen can be obtained since the other components of air (nitrogen and argon) are more volatile and can be removed from the top of the fractionation column as gases.

fraternal twins unidentical twins (dizygotic twins) that develop when two ova are fertilized simultaneously. This occurs when two ova have matured and have been shed simultaneously, and the resultant twins resemble each other only to the same extent as brothers and sisters born at different times.

free energy

free energy (*also called* **Gibb's free energy**) a thermodynamic quantity used in chemistry, which gives a direct criterion of spontaneity of reaction in a reversible process. It is defined by the equation $G = H - TS$, where G is the energy liberated or absorbed, H is the ENTHALPY, S is the ENTROPY, and the system is measured at constant pressure and temperature (T). As a reaction proceeds, reactants form products, and H and S change. These changes, denoted by ΔH and ΔS, result in a change in free energy, ΔG, given by the equation $\Delta G = \Delta H - T\Delta S$. If ΔG is a large negative number, the reaction is spontaneous, and reactants transform almost entirely to products when equilibrium is reached. If ΔG is a large positive number, the reaction is not spontaneous, and reactants do not give significant amounts of products at equilibrium. If ΔG has a small negative or positive value (less than 10 kilojoules), the reaction gives a mixture of both reactants and products in significant amounts at equilibrium.

freeze-drying a process used when dehydrating heat-sensitive substances (such as food and blood plasma) so that they may be preserved without being damaged during the process. The material to be preserved is frozen and placed in a VACUUM. This causes a reduction in pressure, which in turn causes the ice trapped in the material to vaporize, and the water vapour can be removed, producing a dry product. For most solids, the pressure required for vaporization is quite low.

However, ice has an appreciable vapour pressure, which is why snow will disappear in winter even though the temperature is too low for it to melt.

frequency (f) the number of complete wavelengths passing any given reference point on the line of zero disturbance in one second. For example, the frequency of an OSCILLATION, such as a wave, is the number of complete cycles produced in one second, the unit of which is the hertz (Hz). The wave equation is given by $c=f\lambda$, where c = the velocity of the wave and λ is its wavelength. For example, if waves have a wavelength of 2 metres and travel with a velocity of 10 metres per second, then the frequency of the wave motion is 5 Hz.

friction the force that opposes motion and always acts parallel to the surface across which the motion is taking place. Unless a force is exerted to keep an object moving, it will tend to slow down due to the opposing force of friction. Friction can therefore be thought of as a negative force, causing negative acceleration. Frictional forces between two solids or between a solid and a liquid are much greater than those between a solid and air. The hovercraft is a vehicle that exploits this fact by travelling on a cushion of air, thus reducing friction.

fuel any material that, when treated in a particular way, releases energy in the form of heat. The FOSSIL FUELS and organic fuels (e.g. wood and waste material) produce energy by combustion in the presence of oxygen, releasing carbon dioxide and water. Nuclear fuels, such as uranium and plutonium, release large amounts of heat during nuclear reactions when chemical changes occur within the atom (*see* FISSION).

fuller's earth a clay consisting primarily of montmorillonite (a complex silicate of aluminium with water in its structure) used originally to absorb fats from wool (termed "fulling," hence the name). Fuller's earth is now used in the textile industry and in the refining of oils and fats.

function a mathematical term used when there is a relationship between two or more VARIABLES. For example, if two variables are x and y, and there is an associated value for y for every value of x, then y is said to be a function of x. The values of x are termed the domain of such a function, and the range of that function is the term given to the corresponding set of y values.

functional group an arrangement of atoms joined to a carbon skeleton, which gives an organic compound its chemical properties. Compounds with the same functional groups are classed together because of their similar properties. For example, compounds with the functional group NH_2 are classed as amides; those with carbon-carbon double bonds are classed as ALKENES; those with the -OH group are classed as ALCOHOLS.

fundamental particle *see* **elementary particle**.

fungus (*plural* **fungi**) simple unicellular or filamentous plants with no chlorophyll. Fungi cause decay in fabrics, timber and food, and diseases in some plants and animals. Particular fungi are used in brewing, baking, and in the production of ANTIBIOTICS.

fuse a device for maintaining the CURRENT in a CIRCUIT by preventing it from rising too high if a fault should occur. It is simply a thin metal wire of low MELTING POINT so that the heat generated from too high a current melts the wire, causing the circuit to break and the current to fall to zero. Fuses have different ratings according to the thickness of the wire. The fuse rating is the maximum current that can flow through the fuse without causing the circuit to break.

fusion a nuclear reaction in which unstable nuclei combine to create larger, more stable nuclei with the release of vast amounts of energy. For the reaction to occur, the nuclei have to collide, and this requires the nuclei to have very high KINETIC energies to overcome the repulsive forces between them. NUCLEAR FUSION occurs in the hydrogen bomb (fusion bomb), and at temperatures of about 100 million degrees centigrade, the reaction becomes self-sustaining.

G

galaxy (*plural* **galaxies**) the name given to the band of stars, numbering 10^{11} bodies, which includes the sun, and is alternatively called the Milky Way. The galaxy has a spiral structure and is approximately 10^5 LIGHT YEARS across.

gall bladder *see* **bile**.

galvanizing the process by which one type of metal is coated with a thin layer of another, more reactive metal. Galvanizing is performed for the purpose of protection, as the more reactive metal coating will corrode before the underlying metal. For instance, sheets of iron and steel are often coated with the more reactive metal, zinc. Even when the zinc coating becomes damaged, the underlying iron or steel will be protected.

gamete the reproductive cell of an organism. Gametes can be either male or female, and these specialized cells are HAPLOID in number but unite during fertilization, producing a DIPLOID ZYGOTE that later develops into a new organism. In higher animals, the male and female cells are called sperm and ova respectively, whereas in higher plants they are known as pollen grains and egg cells respectively. In some

gas laws

organisms there is essentially one type of gamete that is capable of developing into a new individual without fertilization. These gametes are usually diploid, as in the case of certain lower plant groups, e.g. many forms of algae.

gamma ray a type of ELECTROMAGNETIC radiation released during the radioactive decay of certain nuclei. The rays released are the most penetrative of all radiations, requiring about twenty millimetres of lead to stop them. The gamma rays are useful for sterilizing substances and in the treatment of cancer. They have the shortest wavelength of any wave in the electromagnetic spectrum, i.e. 10^{-10} to 10^{-12} metres.

gas the fluid state of matter capable of indefinite expansion in every direction, due to the relatively few bonds between the atoms or molecules present in the gas. If heat ENERGY is supplied to a gas, it expands to the limits of its containing vessel, exerting a pressure on this vessel that in turn exerts a force back onto the gas.

gas laws the rules that relate to the pressure, temperature and volume of an ideal gas, allowing useful information about a gas to be gained by calculation instead of by experimentation. The laws are termed BOYLE'S LAW, CHARLES' LAW, and the pressure law. The pressure law states that, when a gas is kept in a constant volume, the pressure of that gas will be directly proportional to the temperature. All three laws can be combined in an equation known as the uni-

Geiger tube

versal gas equation, which allows gases to be compared under different temperatures and pressures, i.e. pV = nRT, where p, V and T relate to pressure, volume and temperature respectively, n is the quantity of gas under investigation, and R is the universal molar gas constant, which has the value of 8.314 $JK^{-1}mol^{-1}$.

Geiger tube *or* **Geiger-Müller tube** an instrument, named after the German physicist Hans Geiger (1882-1945), that can detect and measure radiation. The tube contains an inner electrode and a cylindrical outer electrode filled with a gas at low pressure. Any radiation enters the tube through a mica window, causing an electrical pulse to travel between the electrodes. These pulses are detectable when the Geiger tube is connected to an electronic circuit, called a scaler, which records the total radiation in the area in a given time.

gel a jelly-like material resulting from the setting of a COLLOIDal solution. The VISCOSITY is often such that the solution may have properties more like solids than liquids.

gene the chemically complex unit of heredity, found at a specific location on a CHROMOSOME, that is responsible for the transmission of information from one generation to the next. Each gene contributes to a particular characteristic of the organism, and gene size varies according to the characteristic that it codes for. For example, the gene that codes for the HORMONE called insulin, consists of 1700 BASE PAIRS on a DNA molecule.

gene cloning a method of GENETIC ENGINEERING whereby specific genes are extracted from host DNA and introduced into the cell of another host by means of a plasmid VECTOR. All the descendants of the genetically transformed host cell will produce a copy of the gene. The transformed gene is thus said to have been cloned (*see also* CLONE).

gene flow the transfer of genes between populations via the GAMETEs. Gene flow enhances variation in a population as it can lead to a change in the frequency of ALLELEs present within that population. This in turn is a factor that contributes towards EVOLUTION, as the alleles affect the characteristics of an organism. Therefore, gene flow can be advantageous, as it can help an organism inherit new characteristics that may be beneficial to its survival.

generator *see* dynamo.

genetic adaptation *see* adaptation.

genetic engineering the branch of biology that involves the artificial modification of an organism's genetic make-up. The term covers a wide range of techniques, including selective plant and animal breeding, but it is especially associated with two particular techniques: (1) The transfer of DNA from one organism to a different organism in which it would not normally occur. For example, the gene that codes for the human hormone, insulin, has been successfully incorporated into the GENOME of bacterial cells, and the bacteria produce insulin.

genetic recombination

(2) Recombination of DNA between different species in the hope of producing an entirely new species. For instance, cells of the potato and tomato plants, which have had their cell walls removed, have been successfully cultured and made to fuse together using a variety of experimental procedures. Such cells can grow successfully and develop into a new species of plant that has been called the pomato. Although crossing the species barrier is an important breakthrough in the field of genetic engineering, there are strict governmental regulations regarding the release of such species into the environment since the consequences cannot be predicted.

genetic recombination the exchange of genetic material during meiosis, with the effect that the resultant GAMETEs have gene combinations that are not present in either parent. This rearrangement of genes allows for variability in a species, and in each generation an almost infinite variety of new combinations of ALLELEs of different genes are created. Such novel combinations of genes can confer enormous benefits to an organism when conditions change. For example, only a tiny number of a population of locusts have specific combinations of genes that enable them to survive potent pesticides. When such insects reproduce, they produce resistant populations—a major problem in the world of agriculture.

genome the total genetic information stored in the

CHROMOSOMES of an organism. The number of chromosomes is characteristic of that particular species of organism. For instance, a man has 23 pairs of HOMOLOGOUS CHROMOSOMES (containing approximately 50,000 genes), domestic dogs have 39 pairs, and domestic cats have 19 pairs. In each case, one pair of chromosomes constitute the SEX CHROMOSOMES, and the remaining pairs are the AUTOSOMES.

genotype the specific versions of the GENES in an individual's genetic make-up. For instance, there are three possible genotypes for the human albino gene, and it has two allelic forms, dominant A and recessive a. Thus, the three possible genotypes are:
(1) AA (homozygous dominant)
(2) aa (homozygous recessive)
(3) Aa (heterozygous).

geology the scientific study of planet earth. This includes geochemistry, petrology, mineralogy, geophysics, palaeontology, stratigraphy, physical and economic geography.

geometry a major branch of the mathematical sciences, which involves the study of the relative properties of various shapes. For example, the calculations used to determine the size of the angles and the area of a triangle.

geotropism a growth movement, exhibited by plants in response to the force exerted by GRAVITY. Plant roots are termed positively geotropic since they grow downwards, whereas plant shoots generally grow upwards (towards sunlight) thus displaying negative geotropism.

germination the start of growth in a dormant structure, e.g. a seed or spore. Various factors can break seed dormancy, such as specific temperatures, exposure to light, or rupture of the seed coat, all of which depend on the species from which the seed is derived.

gestation period the period from conception to birth in mammals, which is characteristic of the species concerned. For instance, dogs have a gestation period that on average is 63 days, whereas that of the blue whale is 11 months.

gibberellin *see* **hormone**.

Gibb's free energy *see* **free energy**.

gluconeogenesis a major metabolic pathway occurring predominantly in the liver, which synthesizes glucose from non-carbohydrate precursors in conditions of starvation. Glucose is required by red blood cells and is the primary energy source of the brain. However, the glucose reserves present in body fluids are sufficient to meet the body's needs for only about one day. Therefore, gluconeogenesis is very important during longer periods of starvation or during periods of intense muscle exercise. There are three major non-carbohydrate classes that serve as raw materials for gluconeogenesis:

(1) GLYCEROL—derived from fat hydrolysis.
(2) AMINO ACIDS—derived from protein degradation during starvation and from proteins in the diet.

(3) Lactate (lactic acid)—formed by actively contracting muscle when there is an insufficient supply of oxygen. (It is also produced by red blood cells).

glucose the most abundant naturally occurring sugar, which has the general formula $C_6H_{12}O_6$. Glucose is distributed widely in plants and animals and is an important primary energy source, although it is usually converted into polysaccharide carbohydrates, which serve as long-term energy sources. The storage polymers of plants and animals are starch and GLYCOGEN respectively. Other polysaccharides of glucose include chitin and cellulose, which have a structural role and also provide strength.

glycerol a viscous, sweet-smelling alcohol, which has the chemical formula $HOCH_2CH(OH)CH_2OH$. Glucose is widely distributed in plants and animals as it is a component of stored fats. During metabolism, stored fats break down to form the original reactants, glycerol and FATTY ACIDS, while a large amount of energy is released. Glycerol is used commercially to manufacture a wide range of products, including explosives, resins, toilet preparations and foodstuffs.

glycolysis a major metabolic process, occurring in the CYTOPLASM of virtually all living cells, where the breakdown of glucose into simple molecules generates energy in the form of ATP. Each 6-carbon glucose molecule is converted into two 3-carbon pyruvate molecules ($CH_2COCOOH$) in a sequence of ten reactions, giving a net gain of two ATP molecules.

glycosylation

The glycolytic pathway is regulated by several ENZYMES. Although the reactions converting glucose to pyruvate are very similar in all living organisms, the fate of pyruvate is variable. In AEROBIC organisms, pyruvate enters the MITOCHONDRIA, where it is completely oxidized to CO_2 and H_2O in a process known as Kreb's cycle (or CITRIC ACID CYCLE). This cycle, together with glycolysis, liberates 38 molecules of ATP per glucose molecule. However, if there is an insufficient supply of oxygen, e.g. in an actively contracting muscle, FERMENTATION occurs and pyruvate is converted into lactic acid, liberating only 2 ATP molecules per glucose molecule. In some ANAEROBIC organisms, such as YEAST, pyruvate is converted into the alcohol ETHANOL during fermentation, again yielding only 2 ATP molecules. If a cell requires energy, or certain intermediates of the pathway are required for the synthesis of new cellular components, glycolysis proceeds, provided that glucose levels in the blood are abundant. However, when blood-glucose levels are low, e.g. during starvation, glycolysis is inhibited and instead GLUCONEOGENESIS occurs. Glycolysis and gluconoegenesis are reciprocally regulated so that when one process is relatively inactive, the other is highly active.

glycosylation the process by which a carbohydrate is added to an organic compound, for example a protein. Glycosylation may occur in the ENDOPLASMIC RETICULUM or the GOLGI APPARATUS of cells and plays an important role in regulating protein activity.

Golgi apparatus a system of ORGANELLES within the cells of organisms, comprising stacks of flattened sacs that act as the assembly point for the modification, sorting and packaging of large molecules; proteins, for example, undergo GLYCOSYLATION here. Numerous small membrane-bound vesicles surround the Golgi apparatus, and these are thought to transport the modified macromolecules from the Golgi apparatus to the different compartments of the cell. It is named after the Italian physician Camillo Golgi (1844-1926) who discovered its existence.

gonads the reproductive organs of animals, which produce the GAMETES and certain HORMONES. The male and female organs are known as the testes and the ovaries respectively.

gradient a measure of the steepness of a sloping line. A straight line has the equation $y = mx + c$, where x and y are the co-ordinates, c is a constant, and m is the gradient. The steepness of a point on a curve is the gradient of its tangent, which is a straight line drawn to the curve at this point.

gram (g) the basic unit of mass. There are approximately 28g in an ounce, and precisely 1000g in a kilogram.

graph a diagram that represents the relationship between two or more quantities, using dots, lines, bars or curves.

graphite a soft, black, hexagonal variety of carbon

gravity

with a very high melting point that makes it chemically inert. It is a giant structure comprising a series of planes. In any one plane the bonds between the atoms are strong, but bonds between the atoms of different planes are weak. These properties account for the fact that graphite is a good lubricant and conductor of electricity. It is used in the making of pencils and electrodes.

gravity the attractive force that the earth exerts on any body that has mass, tending to cause the body to accelerate towards it. Other planets also exert a force of gravity, but the force is different from that exerted by the earth since it depends on the planet's mass and diameter. The true WEIGHT of any object on earth is really equal to the object's MASS (m) multiplied by the ACCELERATION due to gravity (g), which is 9.8 ms^{-2}. Therefore, although weight and mass are often used synonymously, they are different for scientific purposes. For example, a man with a mass of 80kg will weigh 784 newtons (N) on earth, but on the moon he would weigh only 130N since the force of gravity on the moon is only 1/6th of that on earth. However, his mass is still 80kg and remains constant throughout the universe.

greenhouse effect the phenomenon whereby the earth's surface is warmed by solar radiation. Most of the solar radiation from the sun is absorbed by the earth's surface, which in turn re-emits it as INFRARED RADIATION. However, this radiation becomes trapped in the earth's atmosphere by carbon dioxide (CO_2), water vapour and OZONE, as well as by clouds, and is re-radiated back to earth, causing a rise in global temperature. The concentration of CO_2 in the atmosphere is rising steadily because of mankind's activities (e.g. deforestation and the burning of FOSSIL FUELS), and it is estimated that it will cause the global temperature to rise 1.5-4.5°C in the next fifty years. Such a rise in temperature would be enough to melt a significant amount of polar and other ice, causing the sea level to rise by perhaps as much as a few metres. This could have disastrous consequences for coastal areas, in particular, major port cities like New York.

group the vertical columns of elements in the PERIODIC TABLE. Each group contains elements that have similar properties, including the same number of electrons in the outer energy level (shell). This number is represented by the group number. For example, the alkali metals in group 1 all have one electron in the outer shell, whereas the HALOGENS in group 7 all have 7 electrons in the outer shell.

guanine ($C_5H_5N_5O$) a nitrogenous base component of the nucleic acids, DNA and RNA. It has a PURINE structure with a pair of fused HETEROCYCLIC rings, which contain nitrogen in addition to carbon. In both DNA and RNA, guanine always BASE PAIRS with CYTOSINE, which has a pyrimidine structure. Guanine is also present in many other biologically important molecules.

H

Haber process the industrial process for the production of ammonia (NH_3) by the direct combination of nitrogen and hydrogen in the presence of an iron CATALYST. The process gives a maximum yield (40 per cent) using relatively low temperatures and high pressures. The Haber process is important in industrial chemistry since it is the most economic way to produce ammonia, from which fertilizers are made.

hadron *see* **elementary particle**.

haemoglobin an iron-containing red pigment, which is found within the red blood cells (or ERYTHROCYTES) of vertebrates and which is responsible for the transport of oxygen around the body. In actively metabolizing tissue, e.g. the muscles, haemoglobin exchanges oxygen for carbon dioxide (CO_2), which is then carried in the blood back to the heart and pumped to the lungs, where the haemoglobin loses the CO_2 and regains oxygen.

haemophilia a genetic disorder affecting the blood, in which the lack of a vital BLOOD CLOTTING factor causes abnormally delayed clotting. Haemophilia is exhibited almost exclusively by males, who receive the defective gene from their mothers. A haemophilic female can only arise if a haemophilic male marries a female carrying the gene (extremely rare). There is no known cure for haemophilia, and, when injured, haemophiliacs must rely on blood transfusion to replace the blood loss, which is considerably greater than that lost by a normal individual.

haemostatis *see* **blood clotting**.

half-life (t) the time taken for a radioactive ISOTOPE to lose exactly half of its RADIOACTIVITY. The half-life is constant for a particular isotope, varying from a fraction of a second to millions of years, and is best determined by using a Geiger-counter (GEIGER TUBE). For instance, if an isotope has a half-life of one minute, then the radioactive count will fall by one half in one minute, by one quarter in two minutes, by one eighth in three minutes, and so on.

halide a compound consisting of a HALOGEN and another element. Halides are ionically bonded when formed by electropositive metals, e.g. sodium bromide (NaBr). Halides formed by less electropositive metals and non-metals have COVALENT BONDS.

halogen any of five elements, found in group 7 of the PERIODIC TABLE, that are the extreme form of the non-metals. They exhibit typical non-metal characteristics, existing as COVALENTly BONDed diatomic molecules, e.g. F-F (a fluorine molecule). At room temperature, fluorine and chlorine are gases, bromine is a volatile liquid, and iodine is a volatile solid. The halogens are found in nature as negative IONS in sea water and as salt deposits from dried-up seas.

haploid this term describes a cell nucleus or an organism that possesses only half the normal number of CHROMOSOMES, i.e. a single set of unpaired chromosomes. This is characteristic of the GAMETES and is important at fertilization as it ensures the DIPLOID chromosome number is restored. For example, in a man there are 23 pairs of chromosomes per somatic cell, which is the diploid number, but the gametes possess 23 single chromosomes, which is the haploid number.

hard water water that does not readily form a lather with SOAP. This is due to dissolved compounds of calcium, magnesium and iron. Use of soap produces a scum which is the result of a reaction between the FATTY ACIDS of the soap and the metal ions. The scum is made up of SALTS, which when removed render the water soft. There are two types of hardness. Temporary hardness is created by water passing over carbonate rocks (e.g. limestone or chalk), producing hydrogen carbonates of the metals, which dissolve before the water reaches the mains supply. Boiling the water decomposes the hydrogen carbonates into carbonates (producing kettle fur), and the water becomes soft. Permanent hardness in water is due to metal sulphates, which can be removed by the addition of sodium carbonate. ZEOLITES will remove both types of hardness.

Hardy-Weinberg ratio a law that states that in a large, randomly breeding population, the genetic and allelic frequencies will remain constant from generation to

Hardy-Weinberg ratio

generation. For example, a particular GENE in a population may have a number of ALLELES, one of which is dominant, but this does not necessarily mean that it occurs at a higher frequency than the recessive alleles. If the gene has two alleles, B (dominant) and b (recessive), present at frequencies x and y respectively, then the proportion of the genotypic frequencies would be:

$$BB \quad Bb \quad bb$$
$$x^2 \quad 2xy \quad y^2 = 1.0$$

However, there are certain conditions of stability that must be met for such a genetic equilibrium to occur:

(1) The population must be large, so that allelic frequencies could not be altered by chance alone.
(2) There must be no mutation, or it must occur in equilibrium.
(3) There must be no immigration or emigration, which would alter the genetic frequencies in question.
(4) Mating and reproductive success must be completely random with respect to GENOTYPE.

If all conditions of the Hardy-Weinberg law are met, EVOLUTION could not occur as allelic frequencies would not change. However, evolution does occur because the conditions are never entirely met. For example, the condition of random mating is probably never met in any real population since an organism's genotype almost always influences its choice of a mate and the physical efficiency and frequency of its mating. The condition regarding mutation is probably never met either, since mutations are always occurring, resulting in a slow shift in the allelic frequencies in the population, with the more mutable alleles tending to become less frequent.

heart a hollow, muscular organ that acts as a pump to circulate blood throughout the body. The heart lies in the middle of the chest cavity between the two lungs. It is divided into four chambers, known as the right and left ATRIA, and the right and the left VENTRICLES. In normal persons there is no communication between the right side and the left side of the heart, thus the two sides act as independent pumps, which are connected in series. Starting from the left ventricle, the flow of blood is as follows:

(1) Left ventricle contracts and oxygenated blood is pushed into the AORTA under pressure.

(2) Aorta divides into numerous ARTERIES to supply blood to all parts of the body.

(3) Deoxygenated blood returning from the body is carried by small VEINS, which eventually join up to form two large veins, called the superior VENA CAVA and the inferior vena cava.

(4) These two large veins empty into the right atrium.

(5) The blood passes from the right atrium to the right ventricle via a VALVE.

(6) The right ventricle contracts, pushing blood under pressure into the PULMONARY ARTERY.

(7) The pulmonary artery branches into two, carrying blood to both the right and left lungs.

(8) Within the lungs, gas exchange occurs—carbon dioxide is expelled, and the blood is oxygenated (*see* HAEMOGLOBIN).

(9) The blood flows from the left atrium via a valve into the left ventricle.

heat ENERGY produced by molecular agitation.

heat capacity (C) the quantity of heat required by a substance or material that will raise its temperature by one degree KELVIN (or one Celsius). Thus, heat capacity is measured in joules per Kelvin (JK^{-1}) or joules per Celsius (JC^{-1}). The molar heat capacity of a substance is the heat required that will raise the temperature of one MOLE of the substance by one degree, and the specific heat capacity is the heat capacity per kilogram ($JK^{-1}kg^{-1}$) or gram ($JK^{-1}g^{-1}$).

heat exhaustion a physical state experienced by warm-blooded animals whereby the body's normal cooling processes fail to operate as a result of increasing environmental temperature. Instead, the body's metabolic rate increases, raising the body temperature higher, which in turn raises the metabolic rate even higher, and so on. The symptoms of heat exhaustion are cramp and dizziness, and death ensues when the body temperature reaches about 42°C, which is the upper lethal temperature for the average human being.

hela cell a particular cell variety, discovered in a woman with cervical carcinoma (a form of CANCER) in 1951. These transformed cells are immortal and are used in laboratories worldwide for research purposes.

helium

helium an INERT GAS with a stable NUCLEUS identical to an α particle (*see* ALPHA DECAY). It occurs naturally in small quantities, and, being nonflammable and light, is used for airships and balloons. Helium liquefies below 4K and is the commonest coolant used in CRYOGENICS.

helix a curve in the form of a spiral, which encircles the surface of a cone or cylinder at a constant angle.

hemisphere a half sphere, formed when a plane is passed through the middle of a sphere, cutting it in two.

Henry's law this states that the amount of gas dissolved in a given quantity of liquid is directly proportional to the gas pressure above the solution. Henry's law only applies to oxygen and inert gases, i.e. GROUP 8 of the periodic table. It does not apply to carbon dioxide or ammonia, for example, as both will react with water. Henry's law is represented by the following equation: $c = kP$, where c is MOLARITY, k is the absorption coefficient, and P is the pressure. The temperature must be constant as heat decreases the solubilities of gases in liquids but increases the solubilities of liquids in liquids.

hepatitis a serious disease of the liver, of which there are two types:

(1) Hepatitis A (infectious hepatitis) caused by the hepatitis A virus

(2) Hepatitis B (serum hepatitis) caused by the hepatitis B virus

Both diseases share the same symptoms of fever, nausea and JAUNDICE, but they are transmitted by different routes. Hepatitis A is spread by the oral-faecal route and occurs in people who have poor sanitation and personal hygiene. The virus can be transmitted from person to person in contaminated food or drinking water. Most people exposed to the disease can be protected by PASSIVE immunization, which involves the administration of purified ANTIBODIES from a previously infected individual who has recovered. Hepatitis B is spread through blood products, contaminated syringes and instruments. Susceptible groups include those who require blood or blood products, e.g. haemophiliacs (although any donated blood is normally screened for hepatitis). A significant percentage of hepatitis B sufferers develop cancer, and the virus is thought to be a contributory factor.

heptagon a polygon with seven sides, the interior angles of which add together to give 900°. If the heptagon is regular, then all the sides are equal.

hertz (Hz) the unit of FREQUENCY, equivalent to one cycle per second. For example, if an OSCILLATION has a frequency of 6Hz, this means that six complete cycles occur in one second.

heterocyclic compounds organic compounds forming a ring structure with the additional elements, e.g. oxygen, hydrogen, nitrogen and sulphur.

heterozygote an organism having two different ALLELES of the GENE in question in all somatic cells. For instance, if gene B has two allelic forms, B and b, then the heterozygote will contain both alleles, i.e. Bb, at the appropriate location on a pair of HOMOLOGOUS CHROMOSOMES. Heterozygotes can thus produce two kinds of GAMETES, B and b. One allele of a heterozygote is usually dominant, and the other is usually recessive. The dominant allele is the one that is expressed phenotypically, because it masks the expression of the recessive allele. Dominant alleles are usually denoted by capital letters, while recessive alleles are denoted by lower case letters.

hexagon a polygon with six sides. If the hexagon is regular, then each of the interior angles is 120°.

Hill reaction the light-dependent stage of PHOTOSYNTHESIS, in which illuminated CHLOROPLASTS initiate the photochemical splitting of water. This produces hydrogen atoms (two per water molecule), which are used to reduce carbon dioxide with the formation of carbohydrate in the dark stage of photosynthesis. The light stage also produces ATP, which provides the energy required for carbohydrate synthesis.

histogram a graph that represents the relationship between two variables using parallel bars, but it differs from a bar chart in that the frequency is not represented by the bar height, but by the bar area.

histamine an AMINE released in the body during allergic reactions, and in injured tissues. Release causes dilation of blood vessels, causing a fall in blood pressure.

HIV (*abbreviation for* human immune deficiency virus) the retrovirus thought to be the cause of AIDS.

hole *see* **electric current**.

holography a method of recording and reproducing three-dimensional images using light from a LASER, but without the need for cameras or lenses. The holographic images are generated by two beams of laser light producing interference patterns. A single beam of laser, or coherent, light is split into two. One beam is reflected onto the object and then onto the photographic film or plate. The second reference beam passes straight onto the film. The interference pattern on the film produces a **hologram**. The developed film, when illuminated by coherent light, reproduces the image because the interference patterns break up the light, which reconstructs the original object. Because a screen is not required, the light forms a three-dimensional image in air.

homeostasis the various physiological control mechanisms that operate within an organism to maintain the internal environment at a constant state. For example, homeostasis operates to keep the body temperature of humans within a small, crucial temperature range, independent of the temperature of the external environment, as our metabolic processes would not function in any other temperature range.

homologous pertaining to organs or structures that have evolved from a common ancestor, regardless of their present-day function. For example, the pentadactyl limb is the ancestral form of the quadruped forelimb, and from it evolved the human arm, the fin of cetaceans, and the wings of birds. These structures are therefore said to be homologous. Similarities in homologous structures are best seen in early embryonic development and imply relationships between organisms living today.

homologous chromosomes chromosomes that are identical in their genetic LOCI but can have individual allelic forms (*see* ALLELE) that are not necessarily the same. In DIPLOID organisms, a pair of homologous chromosomes exists in all SOMATIC CELLS, each member of the pair having come from a different parent. During MITOSIS in somatic cells, homologous chromosomes do not associate with each other in any way. However, during MEIOSIS in the formation of the GAMETES, homologous chromosomes join together to form a pair, and exchange of genetic material may occur before the homologous pair separates into two new cells that produce gametes. Therefore, the gametes have only a single set of chromosomes so that at fertilization the diploid number is restored.

homologous series chemical compounds that are related by having the same functional group(s) but formulae that differ by a specific group of atoms. For instance, the ALKENES form an homologous series in which each successive member has an additional CH_2 group, i.e.:

Alkene Series	Molecular Formula
Ethene	C_2H_4
Propene	C_3H_6
Butene	C_4H_8

homozygote an organism that has two identical ALLELES of the GENE in question in all SOMATIC CELLS. For instance, if gene B has two allelic forms, B and b, the homozygote will contain only one allelic type, i.e. either BB or bb, at the appropriate location on a pair of HOMOLOGOUS CHROMOSOMES. Homozygotes can thus only produce one kind of gamete, B or b, and as such are capable of pure breeding. For example, the gene for albinism is RECESSIVE, and any individual that possesses this phenotypic trait will be homozygous for the gene. If two such individuals breed, the resultant offspring will all be albinos.

Hooke's law the physical relationship between the magnitude of the applied force on an elastic material and the resulting extension. The extension must be within the YIELD zone of the material, as any force that goes beyond this will cause permanent deformation. Hooke's law can be represented by the equation $T = kx$, where T is the magnitude of force, k is the spring constant, and x is the displacement of the material.

horizontal the term in mathematics that describes a line that is at right-angles with the vertical and parallel to the horizon.

hormone an organic substance, secreted by living cells of plants and animals, that acts as a chemical messenger within the organism. Hormones act at specific sites, known as "target organs," regulating their activity and eliciting an appropriate response. In animals, hormones are secreted from various ductless glands, which include the pancreas, thyroid and adrenal glands. This hormone-signalling system is collectively known as the ENDOCRINE SYSTEM. These glands secrete hormones directly into the bloodstream, usually in small amounts, where they circulate until they are picked up by appropriate receptors present on the cell membranes of the target organs. These receptors recognize the particular hormone and bind to it, initiating a response. Hormones are also important in plant and seed growth and are found in root tips, buds, and other areas of rapid development. For example, gibberellins are a class of plant hormones involved in initiating responses to light and temperature, the formation of fruit and flowers, and the promotion of seed elongation. Hormone action is constantly regulated by elaborate FEEDBACK MECHANISMS, both within and between cells and organs, that regulate their secretion and breakdown.

hovercraft *see* **friction**.

humidity the amount of water vapour in the earth's atmosphere. The actual mass of water vapour per unit volume of air is known as the absolute humidity and is usually given in kilograms per cubic metre (kgm^{-3}). However, it is useful to use relative humidity, which is the ratio, as a percentage, of the mass of water

hydrolysis

vapour per unit volume of air to the mass of water vapour per unit volume of saturated air at the same temperature.

humus the material in soil that results from the decomposition of animal and vegetable matter. It provides a source of nutrients for plants.

Huntington's chorea *see* **lethal gene**.

hydrocarbon an organic compound that contains carbon and hydrogen only. There are many different hydrocarbon compounds, the most common being the ALKANES, ALKENES, and ALKYNES.

hydrochloric acid an aqueous solution of hydrogen chloride gas, producing a colourless, fuming, corrosive liquid. It will react with metals to form chlorides, liberating hydrogen. It is made by the ELECTROLYSIS of brine, producing hydrogen and chlorine, which are combined, or by the reaction of SULPHURIC ACID with sodium chloride. It has many uses in industry.

hydrogen bomb *see* **nuclear fusion**.

hydrogen peroxide (H_2O_2) a strong oxidizing and bleaching agent usually in the form of a SOLUTION in water. On decomposing it produces water and oxygen, hence it is used as a bleach. It is used industrially and as the oxidizing agent in rocket fuel.

hydrolysis the term used to describe a chemical reaction where the action of water causes the decomposition of another compound and the water itself is decomposed. In salt hydrolysis, the salt dissolves in water, producing a solution that may be neutral, acidic or basic, depending on the relative strengths of the ACID and BASE of the salt. For example, a solution of potassium chloride (KCl) would be neutral, since potassium forms a strong base and chlorine forms a strong acid. In comparison, ammonium chloride (NH_4Cl) gives an acidic solution, since ammonium forms a weak base but chlorine forms a strong acid.

hydrosphere the water that exists on or near to the earth's surface. The main components are water (H_2O), sodium chloride (NaCl) and magnesium chloride ($MgCl_2$). By mass, the major elements are oxygen (almost 86 per cent), hydrogen (10.7 per cent), chlorine (2 per cent) and sodium (1 per cent). Magnesium is the only other element present in significant quantities.

hydroxide a compound derived from water (H_2O) through the replacement of one of the hydrogen atoms by another atom or group, e.g. NaOH, sodium hydroxide. ALKALIS are the hydroxides of metals.

hydroxyl the OH group comprising an oxygen and a hydrogen atom bonded together. In alcohols the OH group occurs in a COVALENTly bonded form.

hypotenuse in a right-angled triangle, this is the longest side, which always faces the right angle.

hysteresis the effect when a physical process lags behind its cause. For example, when a material or body is stressed, the STRESS produces STRAIN. When the stress is removed, the strain is not removed immediately or completely, and a residual strain remains.

I

icosahedron a polyhedron with 20 plane faces. If the faces are equilateral triangles, then the icosahedron is said to be regular.

ideal gas a gas that exists only hypothetically and therefore obeys the GAS LAWS precisely. In an ideal gas, the molecules would occupy negligible space, and the forces of attraction between them would also be negligible. The ideal gas would also show perfect elasticity, since it would not be able to store energy, i.e. the molecules would be able to return to their original dimension after any collisions between neighbouring molecules and their containing vessel.

identical twins the offspring that develop from a single fertilized ovum that splits into two very early during development, producing two separate individuals. Identical twins (also known as monozygotic twins) have precisely the same genetic constitution and are always of the same sex.

Ig *abbreviation for* IMMUNOGLOBIN.

igneous rock one of the three main ROCK types formed by the intrusion of magma at depth in various physical forms or extrusion at the surface as lava flows. Typical rocks are granite, basalt and dolerite.

immiscible the term that describes liquids that cannot be mixed together. Such liquids tend to be polar and non-polar (e.g. water and ether respectively), which, when added together, form two separate layers, with the less dense liquid forming the upper layer. Conversely, two polar liquids, or two non-polar liquids, will mix together and as such are termed miscible.

immune system the defence system within the bodies of vertebrates, which evolved to afford protection against the pathogenic effects of invading micro-organisms and parasites. The immune system confers two types of immunity to an organism:

(1) Innate (or natural) immunity—this is present from birth and is non-specific, operating against almost any substance that threatens the body.

(2) Acquired immunity—this type of immunity is as a consequence of an encounter with a foreign substance, and it is specific against that foreign substance.

immunoglobin (Ig) these are groups of proteins, collectively termed ANTIBODIES, that are produced by specialized cells of the blood, which are called B-CELLS and which can bind to specific ANTIGENS. B-cells are stimulated to divide in the presence of particular antigens, and the resultant daughter cells produce quantities of immunoglobins, which play an important role in the body's IMMUNE SYSTEM.

incandescence the state when a substance or body is at a sufficiently high temperature to emit light, e.g. as in the filament of an electric light bulb.

independent assortment a process suggested by the Austrian monk Gregor Mendel (1822-84) to explain the random distribution of different gene pairs that allows all possible combinations to appear in equal frequency. Independent assortment is also known as MENDEL'S second LAW OF GENETICS.

independent variable in a mathematical expression, this term describes a VARIABLE that can take any value, irrespective of how the other quantities are varied. For example, in the equation $y = 3x + 8$, y is a function of x and is thus the DEPENDENT VARIABLE, while x is the independent variable.

indicator a chemical substance, usually a large complex organic molecule, that is used to detect the presence of other chemicals. Indicators are usually weak ACIDS, the un-ionized form (often written as HA), having a different colour than the ionized form (H and A), due to the negative ion A. The degree of ionization, and thus colour change, depends on the pH of the solution under investigation. In solution, the indicator partially dissociates, i.e. HA —> H and A. Most useful indicators give a distinct colour change over a small pH range, usually about two units. Commonly used indicators include phenophthalein and methyl orange.

inert gas the elements comprising group 8 of the PERIODIC TABLE. The inert gases are helium, neon, argon, krypton, xenon and radon. They are unreactive monatomic gases, which make up about 1 per cent of

inertia

air by volume, argon being the most abundant. The outer shell comprises eight electrons, which gives these elements their stable configuration since they will not readily lose or gain electrons. Xenon, neon and krypton are extracted from liquid air by FRACTIONAL DISTLLATION, and are used commercially in fluorescent lamps. Helium occurs in natural gas deposits, which is its principal source, and is important in low-temperature research, since it has a boiling point lower than any other substance.

inertia the property of a body that causes it to oppose any change in its present state of motion. Thus, unless a body is acted upon by an external force, it will remain at rest or continue moving at a constant speed in a straight line. The first of NEWTON'S LAWS OF MOTION states that the mass of an object gives a direct measure of its resistance to changing its direction of motion, i.e. its inertia.

infinity (∞) the term used to describe a number or quantity with a value too great to be measured. For example, outer space is regarded as boundless and is therefore described as infinite. Sometimes the symbol is written as $-\infty$, negative infinity, if the value is so small as to be incalculable and insignificant for all practical purposes. Such a value is described as being infinitesimal.

infrared radiation electromagnetic radiation with wavelengths between those of visible light and microwaves, i.e. from about 0.75 µm to 1 mm. Such radiation can penetrate fog, and, by using special photographic plates, details invisible to the naked eye may be rendered visible (*see also* ELECTROMAGNETIC WAVES).

inorganic chemistry the division of chemistry concerned with ELEMENTs and their compounds other than carbon. It thus excludes organic compounds such as alcohols, esters, and hydrocarbons. However, simple carbon compounds, such as metal carbonates, oxides of carbon, and the physical and chemical properties of the element carbon, are usually included.

insoluble a term used to describe substances that will not dissolve in a SOLVENT, or that will dissolve to only a very limited extent. As the term is not precise, different parameters may be set, but generally a substance is said to be insoluble if it will dissolve only to the extent of 0.1 gram or less per 100 millilitres of solvent. A slightly soluble substance has a SOLUBILITY of between 0.0 and 1 gram per 100 ml of solvent, whereas a soluble substance has a solubility of 1 gram or more per 100 ml of solvent.

insulator a substance that is good at retaining heat and preventing electric flow since it is a poor CONDUCTOR. This is usually due to the fact that insulators have few mobile ELECTRONS because all the electrons are strongly attracted to the PROTONS in the atoms of the insulator. Materials commonly used as insulators include plastics, ceramics, rubber, and many other non-metals.

insulin a pancreatic HORMONE that initiates glucose uptake by body cells and thus controls glucose levels in the blood. Insulin functions by stimulating certain PROTEINS found on the surface of cells within the vertebrate body to take up glucose, which would otherwise be unable to enter cells, as it is very hydrophobic. *Diabetes mellitus* is a condition in which the blood contains excessively high glucose levels due to an under-production of insulin, and the excess glucose is excreted in the urine. This condition can prove fatal, but sufferers can be successfully treated by insulin therapy.

integer any number that is not a fraction but a whole number. Integers can have positive or negative values or be zero. The following can all be thought of as integers: -4, -3, -2, -1, 0, 1, 2, 3 and 4.

integration a branch of CALCULUS that is used to evaluate an area under the curve of a particular function. There are various methods of integration, ranging from simple formulae, which can be thought of as the inverse of differential formulae, to the more complex formulae that are used to find the inverse of irregular functions.

interference the interaction between two or more waves passing through a medium at the same place at the same time, resulting in specific patterns. Constructive interference occurs between waves that are in phase, i.e. their crests or troughs overlap, producing a wave of greater AMPLITUDE. A wave with maximum amplitude is produced when two waves are precisely in phase. Destructive interference occurs when waves are out of phase, i.e. where a crest and a trough overlap, producing a wave with reduced amplitude. If two waves are exactly out of phase then they cancel each other out. For example, with light waves, interference may produce alternate bright and dark bands—the brightest bands correspond to waves with maximum amplitude (constructive interference) and the darkest bands correspond to waves with minimum amplitude (destructive interference). Similarly with sound, intervals of silence and increased volume correspond to waves displaying destructive and constructive interference respectively. Interference patterns can also be detected in radio waves and are the cause of distortions in reception.

interphase *see* **mitosis**.

in vitro a term applied to experiments or techniques undertaken in the laboratory, where biological or biochemical processes are carried out "in glass."

in vivo biological and biochemical processes that occur in a living organism or cell.

ion an ATOM or MOLECULE with a positive charge due to electron loss (a CATION) or a negative charge due to electron gain (an ANION). The process of producing ions is known as ionization, and it can occur in a number of ways, including a molecule dissociating into ions when it is added to a solution, or the formation of ions by bombarding atoms with radiation.

ion exchange the exchange of IONs of like charge between a solution and a highly insoluble solid. The solid (ion-exchanger) consists of an open molecular structure containing active ions that exchange reversibly with other ions in the surrounding solution without any physical changes occurring in the material. For example, ion exchange is also used to soften hard water by removing calcium ions. The water is passed through a column containing an exchange resin that contains sodium ions, and these are exchanged for the calcium ions in the hard water, leaving the water calcium-free.

iron (Fe) a metallic ELEMENT in group 8 of the PERIODIC TABLE. It occurs naturally as magnetite (Fe_3O_4), haematite (Fe_2O_3), limonite ($FeO(OH)nH_2O$), pyrite (FeS_2) and siderite ($FeCO_3$). It is extracted from its ores by the blast furnace process. Often in combination with other elements, it is the most widely used of all metals.

irrational number a quantity that cannot be expressed as a fraction and can only be approximately expressed as a decimal. The square root of 2 ($\sqrt{2}$) is an irrational number with the approximate decimal value of 1.414. Irrational numbers together with the set of RATIONAL NUMBERS comprise the set of REAL NUMBERS.

isobar a line used to join points of equal atmospheric pressure on a weather map at a given time. If there is a great change in pressure over a small area then the change in weather is more apparent, and this is shown on a weather map by closely drawn isobars.

isomer a chemical compound that has the same molecular formula as another chemical compound but differs in the arrangement of the constituent atoms. Isomers are studied STEREOCHEMISTRY, which is concerned with the spatial aspects of the structure of molecules. All isomers fall into one of two groups:

(1) Structural isomers—isomers that differ in the way their atoms are bonded to each other.

(2) Stereoisomers—isomers that have atoms that are bonded in the same way but differ in the way they are arranged in space. Isomers usually have different chemical properties and different physical properties (melting point, boiling point, density, etc).

isosceles the term that describes a triangle that has two angles and sides of equal magnitude. It also describes a type of TRAPEZIUM, where the two non-parallel sides are equal.

isotope an atom that differs from other atoms of the same element due to a different number of NEUTRONS within its nucleus. As isotopes still have the same number of PROTONS, their ATOMIC NUMBER is unchanged, but the varying number of neutrons affects their MASS NUMBER. Most elements exist naturally as a mixture of isotopes but can be separated (in a MASS SPECTROMETER) due to the fact that they have slightly different physical properties.

isotropy the feature whereby the properties of a substance do not vary with direction (the opposite of ANISOTROPY).

J

jaundice a condition characterized by the unusual presence of bile pigment circulating in the blood. Jaundice is caused by the bile produced in the liver passing into the circulation instead of the intestines because of some form of obstruction. The symptoms of jaundice include a yellowing of the skin and the whites of the eyes.

joule the unit of all ENERGY measurements. It is the mechanical equivalent of heat, and one joule (J) is equal to a force of one NEWTON moving one metre, i.e. 1J = 1Nm. It is named after James Prescott Joule (1818-1889), a British physicist who investigated the relationship between mechanical, electrical and heat energy, and, from such investigations, proposed the first law of THERMODYNAMICS, the conservation of energy.

K

karotype the number, shapes and sizes of the chromosomes within the cells of an organism. Every organism has a karyotype that is characteristic of its own species, but different species have very different karotypes. For example, all normal human females have 22 pairs of DIPLOID chromosomes with similar shape and size, but all female horses have 32 pairs of diploid chromosomes with their own unique shape and size.

Kelvin scale the unit of temperature (K) based on the temperature scale devised by the British physicist Lord William Kelvin (1824-1907). The Kelvin scale has positive values only with the lowest possible unit of 0K, which is equal to -273.15°C or -459.67°F.

kerosene (*or* **kerosine**) a thin oil that is one of many products obtained during the FRACTIONAL DISTILLATION of PETROLEUM. Kerosene is used as fuel for jet engine aircraft.

ketone an organic compound that contains a C=O (carbonyl) group within the compound as opposed to either end of the compound. There are many forms of ketones, and their physical and chemical properties differ due to the presence of alkyl groups ($-CH_3$) or aryl groups ($-C_6H_5$) within the ketone molecule. Ketones can be detected within the bodies of humans when fat stores are metabolized to provide energy if food intake is insufficient. If these accumulate within the blood, the undernourished person will experience headaches and nausea. The presence of ketones in urine is called ketonuria.

kilobyte (KB) *see* **byte**.

kilocalorie a unit of heat used to express the energy value of food. One kilocalorie is the heat needed to raise the temperature of one kilogramme of water by 1°C. It is estimated that the average person needs 3000 kilocalories per day, but this requirement will vary with the age, height, weight, sex and activity of the individual.

kilogram a unit of mass (kg) that is equal to the international prototype made of platinum and iridium stored in the French town of Sèvres.

kinesis the response of an organism to a particular stimulus in which the response is proportional to the intensity of the stimulation.

kinetic energy the energy possessed by a moving body by virtue of its mass (m) and velocity (v). The kinetic energy (Ek) of any moving body can be determined using the following equation:

$$Ek = \tfrac{1}{2}mv^2$$

(the energy is in joules if m is kg and v ms^{-1})

As the kinetic theory of matter states that all matter consists of moving particles, it holds that all particles must possess some amount of kinetic energy, which will increase or decrease with the surrounding temperature (*see* EVAPORATION).

Klinefelter's syndrome a condition in which human males have the abnormal GENOTYPE of XXY rather than normal XY. This produces recognizable characteristics within the affected male, such as the development of breasts and smaller testes (resulting in reduced fertility). Klinefelter's syndrome occurs in approximately one in a thousand male births and is caused by nondisjunction of sex chromosomes during MEIOSIS.

kneejerk reflex a complex neural pathway in humans in which a blow just beneath the kneecap (patella) results in a rapid extension of the leg. The knee jerk is a response by the central nervous system to stimulation of sensory NEURONS located at the front of the thigh. The response leads to the contraction of one muscle (quadriceps, at the front of the thigh) and the inhibition of the contraction of another muscle (biceps, at the back). As the quadriceps muscle is the extensor, the leg straightens but is prevented from bending as the biceps muscle, which is the flexor, is inhibited from contracting.

knot a unit of nautical speed equal to one NAUTICAL MILE (1.15 statute miles or 1.85 kilometres) per hour. The term knot originates from the period when sailors calculated their speed by using a rope with equally spaced tied knots, attached to a heavy log trailing

Korsakoff's syndrome

behind the ship. The regular space between knots was measured at 47 feet, 3 inches, which is 14.4 metres.
Korsakoff's syndrome (*or* **psychosis**) a neurological disorder first described by a Russian neuropsychiatrist called Sergei Korsakoff (1854-1900). The condition is characterized by gross defects in memory for recent events, disorientation, and no appreciation of time. Patients with Korsakoff's syndrome are unaware that there is a problem and are liable to confabulate. Although it can result from lack of vitamins or lead and manganese poisoning, Korsakoff's syndrome most commonly occurs as a complication of chronic alcoholism. It is caused by a dietary deficiency of vitamin B1 (thiamine) which is needed for the conversion of carbohydrate to glucose. The syndrome has been invaluable in neuropsychology, as it has helped to identify the brain regions involved in the memory processes of recall and recognition.
Kreb's cycle *see* **citric acid cycle**.

L

labelled compound a compound used in radioactive tracing, where an atom of the compound is replaced by a radioactive ISOTOPE, which can be followed through a biological or physical system by means of the RADIATION it emits.

laevorotatory compound any substance that, when in crystal or solution form, has an optically active property that rotates the plane of polarized light to the left (anticlockwise).

lanthanides (*otherwise known as the* **rare earth elements**) these elements, from lanthanum (La) to lutetium (Lu), have much in common, chemically, with the scandium group (group 3B of the PERIODIC TABLE). The properties of these metals are very similar, and the lanthanides and yttrium (symbol Y) all occur together and are separated by CHROMOTOGRAPHY. The elements are reactive with the heavier ones resembling calcium, while scandium is similar to ALUMINIUM.

laser (*acronym for* Light Amplification by Stimulated Emission of Radiation) a device that produces a powerful and narrow monochromatic beam of light. Lasers are used extensively in electronic engineering, fibre

latent heat

optic communications, HOLOGRAPHY, and in certain surgical operations.

latent heat the measurement of heat ENERGY involved when a substance changes state. While the change of state is occurring, the gas, liquid or solid will remain at constant temperature, independent of the quantity of heat applied to the substance (an increase in heat will just speed up the process). The specific latent heat of fusion is the heat needed to change one kilogram of a solid into its liquid state at the MELTING POINT for that solid. For example, the specific latent heat of fusion for pure, frozen water (ice) at 273K (0°C) is 334 kJkg^{-1}. The specific latent heat of vaporization is the heat needed to change one kilogram of the pure liquid to vapour at its boiling point. In the case of pure water again, at its boiling point of 373K (100°C), 2260kJkg^{-1} is the specific latent heat of vaporization needed to change water into steam.

latitude the angular distance of a particular point on the earth's surface relative to the earth's equator. Latitude is measured in degrees corresponding to the angle of incident light from a specific star, e.g. the sun, above the horizon at a given time and is described as being north or south of the equator. On a world globe, lines of latitude are represented by parallel, horizontal lines.

Laurent series a mathematical expression that is particularly useful in the analysis of an area between two CONCENTRIC circles. The given function is expressed as both a positive and a negative infinite POWER SERIES, using the following formula:

$$f(z) = \sum_{n=-\infty}^{\infty} a_n (z-a)^n$$

law of conservation *see* **energy, thermodynamics**.

LD50 the amount of a toxic substance, which, when applied in a specific manner, will kill 50 per cent of a large number of individuals within a species.

LDL-receptor *see* **cholesterol**.

Le Chatelier's principle a statement relevant to chemical reactions, which predicts that if the conditions of a system in EQUILIBRIUM are changed, the system will attempt to reduce the enforced change by shifting equilibrium.

legionnaires' disease an infectious disease caused by the bacterium *Legionella pneumophila*, which inhabits surface soil and water. It has also been traced in water used in air-conditioning cooling towers. The main source of infection is inhalation of air or water carrying the bacteria, and so far there is no evidence that it is transmitted from an infected to a non-infected individual. Legionnaires' disease is really a form of pneumonia, and thus its symptoms include shortness of breath, coughing, shivering and a rise in body temperature. Healthy individuals should fully recover from infection if treated with the antibiotic called erythromycin.

Leibniz's theorem

Leibniz's theorem a mathematical formula that states that the nth derivative [uv]^n of the product of the function u and the function v can be calculated using the following:

$$\sum_{i=0}^{n} \binom{r}{i} u^{(i)} v^{(n-1)}$$

This allows the limit of the first n terms to be reached. The derivative of the product of two functions, u and v, is obtained using the PRODUCT RULE. Gottfried Wilhelm Leibniz (1646-1716) was a philosopher and scientist born in the German town of Leipzig. He devised differential and integral CALCULUS and developed certain aspects of DYNAMICS. He contributed to the development of many other mathematical, philosophical and geological theories, such as the BINARY number system and the theory that present-day earth evolved from a molten origin.

Lenz's law the principle devised by the German physicist H.F.E. Lenz (1804-1865) to devise the direction of an induced current during ELECTROMAGNETIC INDUCTION. Lenz's law states that an induced current will flow in a direction that will oppose the change that induced the current. Lenz's law explains the phenomenon of a suspended closed coil of wire always repelling an approaching magnet but attracting a departing magnet, independent of the orientation of the magnet.

leprosy an infectious disease that affects the skin, nerves and mucous membranes of the patient. The symptoms of leprosy include severe lesions of the skin and destruction of nerves, which can lead to disfigurements such as wrist-drop and claw-foot. Leprosy is caused by the airborne bacterium, *Mycobacterium lepra*, and, fortunately, is not highly contagious as transmission involves direct contact with this bacterium. The likeliest source of infection, therefore, arises from the nasal secretions (swarming with bacteria) of patients and not from the popular misconception of touching the skin of an infected individual. Leprosy is curable, and the treatment, using sulphone drugs, has the beneficial side-effect of making the patient non-infectious even if he or she is not completely cured. Although the incidence of leprosy was once worldwide, it is now mostly confined to tropical and subtropical regions.

lepton *see* **elementary particle**.

lethal gene a gene that, if expressed, will cause the death of the individual. The fatal effect of the expressed gene usually occurs in the prenatal developmental stage of the individual, i.e. the embryonic stage for animals and the pupal stage for insects. Although most examples of lethal mutants fail to survive to adulthood, there is one well-researched genetic disorder, called Huntington's chorea, which does not usually affect the individual until middle age. Huntington's chorea is caused by a single dominant

leucocyte

gene, and thus half the children of an affected parent will inherit the genetic disorder, although fortunately this disease is rare.

leucocyte a large, colourless cell formed in the bone marrow and subsequently found in the blood of all normal vertebrates. It is commonly known as a white blood cell and plays an important role in the IMMUNE SYSTEM of an individual. Leucocytes are produced in the bone marrow, spleen, thymus and lymph nodes of the body and can be classified into the following three groups in order of decreasing constituency of leucocytes:

Group	%	Functions
Granulocyte	70	Helps combat bacterial and viral infection and may also be involved in allergies.
LYMPHOCYTE	25	Destroys any foreign bodies either directly (T-CELLS) or indirectly by producing antibodies (B-CELLS).
Monocyte	5	Ingests bacteria and foreign bodies by the mechanism called PHAGOCYTOSIS.

leucoplast a colourless object that contains starch and is found in some plant cells. If a leucoplast contains the pigment CHLOROPHYLL it may develop into a CHLOROPLAST.

leukaemia a cancerous disease in which there is an uncontrolled proliferation of white blood cells (LEU-

COCYTES) in the bone marrow. The white blood cells fail to mature to adult cells and thus they cannot function as an important part of the defence system against infections. Although the definite cause of leukaemia is as yet unknown, there is growing suspicion that certain viruses may cause it and that perhaps there is an hereditary component. Unfortunately, leukaemia is not a curable disease, but there are methods effective in suppressing the reproduction of white blood cells—radiotherapy and, more commonly, chemotherapy. These methods bring the disease under control and thus help prolong the patient's life.

Lewis acid in a chemical reaction, any substance accepting an ELECTRON PAIR.

Lewis base in a chemical reaction, any substance donating an ELECTRON PAIR.

ligand any MOLECULE or ATOM capable of forming a bond with another molecule (usually a metallic CATION) by donating an ELECTRON PAIR to form a complex ION. In biological terms, ligand refers to any molecule capable of binding with a specific ANTIBODY.

lightning the discharge of high-voltage electricity between a cloud and its base, and between the base of the cloud and the earth. One flash of lightning actually consists of several separate strokes that follow each other at intervals of fractions of a second (too fast for the human eye to detect) and occurs when the strength of the ELECTRIC FIELDS becomes great enough to overcome the RESISTANCE of the intervening air. The

light reactions

clap of thunder that can be heard during thunderstorms is a result of the expansion of the intervening air.

light reactions the biochemical processes that generate ATP, oxygen and a reduced coenzyme called NADPH during PHOTOSYNTHESIS in the presence of light. The light-dependent reactions occur in the inner membranes of CHLOROPLASTS and require water and several forms of the pigment CHLOROPHYLL. Two of the products of the light reactions, ATP and NADPH, enter the CALVIN CYCLE, whereas the third product, OXYGEN, is released by the plant.

light wave *see* **electromagnetic waves**.

light year a measure of the distance travelled by light in one year, which is approximately 9.467×10^{12} kilometres.

limiting reactant any substance that limits the quantity of the product obtained during a chemical reaction. The limiting reactant can be identified using the chemical equation of a reaction as it will be the smallest quantity in comparison to the other reactants and products.

linear equation a mathematical term used to describe any equation containing VARIABLES that are not raised to any power although the variables may have a COEFFICIENT. For example, the following equations are linear combinations of a,b,c and x,y,z respectively:

$$a + 2b = 4c \text{ and } 2x + 3y + z = 0.$$

linkage the association between two or more GENES situated on the same CHROMOSOME. The genes tend to be inherited together, thus the parental gametes, which after FERTILIZATION eventually form the offspring, will not have undergone normal GENETIC RECOMBINATION to generate new combinations of genes. As the distance separating two genes decreases, the chance of these two genes becoming separated during CROSSING-OVER also decreases, thus increasing the chance that these genes are linked to be inherited as one segment rather than separate genes.

lipase any enzyme capable of breaking down fat to form FATTY ACIDS and GLYCEROL. Lipases function in alkaline conditions and are most abundant in the pancreatic secretions during digestion.

lipids an all-embracing term for oils, fats, waxes and related products in living tissues. They are ESTERS of FATTY ACIDS and form three groups: simple lipids, including fats, oils and waxes; compound lipids, which includes PHOSPHOLIPIDS; and derived lipids which includes STEROIDS.

lipoprotein (LDL) any protein that has a FATTY ACID as a side chain. Lipoproteins have significant importance in certain biological processes as they function as a transport mechanism for essential molecules. For example, as CHOLESTEROL is extremely hydrophobic, there would be no method of transporting it to its target body tissues. This problem is solved by low-density lipoprotein surrounding the cholesterol molecule and forming a hydrophilic molecule that can be transported by body fluids.

liquefaction of gases a gas may be turned into its liquid form by cooling below its critical temperature (the temperature above which a gas cannot be liquefied by PRESSURE alone). In addition, pressure may be required. For gases such as oxygen, helium and nitrogen, low temperatures are used.

liquid a fluid state of matter that has no definite shape and will aquire the shape of its containing vessel as it has little resistance to external forces. A liquid can be regarded as having more KINETIC ENERGY than a SOLID but less kinetic energy than a GAS. It is considered that the average kinetic energy will increase as the temperature of the liquid rises.

liquid crystal one of certain liquids that show regions of aligned molecules that are similar to crystals. The application of a current disrupts the molecules sufficiently to darken the liquid and form, for example, characters on a display.

litre a unit of volume given the symbol l and equal to 1000 cubic centimetres, i.e. $1l = 1000 cm^3 = 1 dm^3$. One gallon is approximately 4.5 litres.

local gravitational constant the quantity given to the ACCELERATION of any object near to sea level at any point on earth. This acceleration is a result of gravity and is given the symbol g. It is calculated using the following:

where $$g = \gamma MR^{-2}$$

= universal gravitational constant, M = mass of the earth, R = radius of the earth at that point. At the

longitude

North Pole, for example, g = 9.8321 ms^{-2}, and at the equator g = 9.7801 ms^{-2}.

locus (*plural* **loci**) the set of specific points that either satisfies or is determined by a certain mathematical condition. The locus can be thought of as tracing the path of a moving point relative to another fixed point. For example, a circle is a locus of a point that moves in such a way that the distance (radius of circle) between the moving point and the fixed point (centre of the circle) is constant. In biology, the locus is the name given to the region of a CHROMOSOME occupied by a particular GENE.

logarithm (*abbreviation* **log**) a mathematical FUNCTION that was first introduced as a labour-saving device when dealing with the multiplication and division of large numbers. However, the modern calculator has reduced this need for logarithms, and now a logarithm is the power of a number to a specified base written in the form $\log_a x = n$ (a=base). There are two forms of logarithm:

(1) Common logarithm—has base 10, $\log_{10} x$

(2) Natural or Napierian logarithm—has base e, ln x (where e is EXPONENTIAL FUNCTION).

longitude the angular distance of a given point on the earth's surface relative to the Greenwich meridian (a theoretical line that runs through the North and South Poles as well as Greenwich, England). Longitude is measured in degrees east and west of the Greenwich meridian (this is given an arbitrary value of 0°). By international agreement, the world is divided into 24 longitudinal zones, each 15° in width, starting and finishing at Greenwich (0°), to be able to relate the different times in places throughout the world. Thus, any place in the zone centred at 15° east of Greenwich is one hour ahead of Greenwich, whereas any place in the zone centred at 15° west is one hour behind Greenwich.

longitudinal wave the classification for a wave that is produced when the vibrations occur in the same direction as the direction of travel for that wave. The most well-known example of a longitudinal wave is a SOUND wave, which is propagated by displacement of air particles, causing areas of high density (called compressions) and areas of low density (called rarefactions). The WAVELENGTH of a longitudinal wave is the combined length of a single compression and rarefaction. For the FREQUENCY and speed of a longitudinal wave, *see* WAVE.

loudspeaker a TRANSDUCER that converts electric current into sounds through the vibration of a paper cone or diaphragm.

lubricant any substance that, when applied between surfaces, will cause a decrease in FRICTION. Some common lubricants include oil and graphite dust.

luminescence the emission of light by a living organism that is not a consequence of raising the body temperature of the organism. For luminescence to occur, the cells of the organism must contain the protein

lymph a colourless, watery fluid that surrounds the body cells of vertebrates. It circulates in the LYMPHATIC SYSTEM but is moved by the action of muscles, as opposed to the contraction of the heart. Lymph consists of 95 per cent water but contains protein, sugar, salts, and LEUCOCYTES, and transports fat from the gut wall during digestion.

lymphatic system a network of tubules found in all parts of a vertebrate's body except, if present, the central nervous system. The tubules drain the body fluid, LYMPH, from the tissue spaces, and as they gradually unite to form larger vessels, the lymph vessels finally drain into two major lymphatic vessels that empty into veins at the base of the neck. The system consists of vessels and nodes, where the fluid is filtered, bacteria and other foreign bodies are destroyed, and LYMPHOCYTEs enter the lymphatic system. The lymphatic system is not only an essential part of the IMMUNE SYSTEM, but also carries excess protein and water to the blood and transports digested fats.

lymphocyte a cell that is produced in the bone marrow and is an essential component of the IMMUNE SYSTEM as it will differentiate into either a B-CELL or a T-CELL. Lymphocytes are a form of white blood cell (LEUCOCYTE), which collect in the lymph nodes of the lymphatic system and defend the body against foreign

lysis

bodies and bacterial infections. Lymphocytes are also present in blood but in a smaller percentage.

lysis the destruction of cells, commonly BLOOD CELLS, by antibodies called lysins. More generally, the destruction of cells or tissues by pathological processes, e.g. autolysis, where cells are broken down by enzymes produced in the cells undergoing breakdown.

M

Mach number the speed of a body expressed as a ratio with the SPEED OF SOUND. Mach 1 is sonic speed, and thus subsonic and supersonic are below and above unity, respectively.

Maclaurin's formula a mathematical formula devised by Colin Maclaurin, an 18th-century Scottish polymath, used to expand certain functions around zero. When a function f(x) is infinitely differentiable and has real values, then it can be estimated as the sum of f(0) and the first few terms of the series using the following:

$$f(x) = f(o) + xf'(o) + \frac{x^2}{2!}f''(o) + \cdots + \frac{x^r}{r!}f^{(r)}(o) + \cdots$$

macrophage a specialized cell that forms part of the IMMUNE SYSTEM of vertebrates. Macrophages are derived from MONOCYTES in the blood system and can move to infected or inflamed areas of the body using PSEUDOPODIA. In such areas, they ingest and degrade broken cells and other debris including microbes by means of PHAGOCYTOSIS.

Magellanic clouds two separate GALAXIES, detached from the Milky Way, which appear, from the southern hemisphere, as patches of light. They are approximately 180,000 LIGHT YEARS away and contain a few thousand million stars.

magma the fluid rock beneath the earth's surface, which solidifies to form IGNEOUS ROCKS. During volcanic eruptions, the lava extruded at the earth's surface is not necessarily the same in composition as the magma that arises to form lava, since the magma may have lost some of its gaseous elements and some solids of the magma may have crystallized.

magnetic bubble a portion of computer memory which consists of a small region in a material such as garnet (a silicate mineral), which is magnetized in one direction. Slices of this material placed on a SUBSTRATE produce a magnetic CHIP that under a magnetic field produces magnetic bubbles. Information can be stored on the chip, which may contain up to one million bubbles in 20mm^2, in BINARY form, through the presence or absence of a bubble in a specific location.

magnetic field the region of space in which a magnetic body exerts its force. Magnetic fields are produced by moving charged particles and represent a force with a definite direction. There is a magnetic field covering all of the earth's surface, which is believed to be a result of the iron-nickel core.

magnitude the absolute value or length of a physical or mathematical quantity.

malaria an infectious disease caused by certain parasites in the blood of the victim. Malaria is transmitted by an infected mosquito biting a human, thus injecting the parasite from the salivary gland of the mosquito into the bloodstream of the human. After the parasites have developed in the victim's liver, they are released into the bloodstream and attack red blood cells. Early symptoms of malaria include headaches, body aches and chills. As the disease progresses, malarial attacks are frequent and cause sickness, dizziness and sometimes delirium in the victim; the attacks seem to coincide with the bursting of infected red blood cells. Fortunately, there are many highly effective drugs for treating malaria.

malleability the property of metals and alloys that enables them to be changed in shape by hammering or rolling (or similar processes) into thin sheets.

map unit the region between two gene pairs on the same CHROMOSOME, where one per cent of the possible products during MEIOSIS are recombinant.

mass the measure of the quantity of matter that a substance possesses. Mass is measured in grams (g) or kilograms (kg).

mass number (A) the total number of PROTONS and NEUTRONS in the nucleus of any atom. The mass number therefore approximates to the RELATIVE ATOMIC MASS of an atom.

mass spectrometer the machine used to detect the various types of ISOTOPES found in an element. The mass spectrometer bombards molecules with high-energy electrons, creating smaller positive IONS and neutral fragments. These positive ions are then deflected, using a MAGNETIC FIELD that separates them according to their MASS. The ions finally pass through a slit to the ion collector, and each peak of the printed chart corresponds to a particular ion and its mass. The mass spectrometer also provides information on the relative amount of each isotope present as well as the exact mass of the various isotopes.

mast cell a large, blood-borne cell that has a fast-acting role in the IMMUNE SYSTEM. In allergies, mast cells will be triggered to release histamine when the IMMUNOGLOBIN, IgE, has already attached itself to the foreign body that has entered the body.

matrix (*plural* **matrices**) an array of elements set out in rows and columns. Matrices are a mathematical tool useful in, for example, solving the transformation of co-ordinates. The order of a matrix refers to the number of rows and columns, e.g.:

$$A = (4\ 1\ 6) \qquad B = \begin{pmatrix} 2 & 6 \\ 1 & 4 \end{pmatrix}$$

Matrix A has one row, three columns, and matrix B has two rows, two columns. Only matrices of the same order can be added or subtracted. Matrices are only compatible for multiplication if the number of columns of the first equals the number of rows of the second. To multiply, therefore, the row of one matrix is multiplied by the column of the other matrix, and the products are added.

matter any substance that occupies space and has

MASS: the material of which the universe is made.

megabyte (MB) *see* **byte**.

meiosis a chromosomal division that produces the germ cells (GAMETES in animals and some plants, sexual spores in fungi). Meiosis involves the same stages as MITOSIS, but each stage occurs twice, and, as a consequence, four HAPLOID cells are produced from one DIPLOID cell. Meiosis is an extremely important aspect of sexual reproduction, as the production of haploid cells ensures that during FERTILIZATION the chromosomal number is constant for every generation. It also gives rise to genetic variation in the daughter haploid cells by the rearrangement and GENETIC RECOMBINATION if genetic variability already exists in the parent diploid cell.

melanin a dark pigment responsible for colouring the skin and hair of many animals, including humans. Differences in skin colour are due to variations in the distribution of melanin in the skin and not to differences in the number of cells, **melanocytes**, that produce the pigment.

melting point the temperature at which a substance is in a state of equilibrium between the solid and liquid states, e.g. ice/water. At 1 atmosphere pressure, the melting point of a pure substance is a constant and is the same as the freezing point of that substance. At constant pressure, the melting point of a substance is lowered if it contains impurities (the reason for adding salt to make ice melt). Although an increase in pres-

membrane potential

sure lowers the melting point of ice, in most substances increased pressure will raise their melting points.

membrane potential the difference in electric potential between the inside and the outside of the plasma membrane of all animal cells. Membrane potentials exist due to the different ionic concentrations within and outwith the cell, and also the selective permeability of the plasma membrane to specific ions (notably potassium [K^+], sodium [Na^+] and chloride [Cl^-] ions). When measured using a microelectrode, the resting potential of most muscle and nerve cells is -60mV on the inside of the plasma membrane.

Mendel's laws of genetics laws of heredity deduced by Gregor Mendel (1822-1884), an Austrian monk, who discovered 16 basic rules experiments with generations of pea plants. Mendel discovered that a trait, such as flower colour or plant height, had two factors (hereditary units) and that these factors do not blend but can be either dominant or recessive. Without knowledge of genes or cell division, he developed the following laws of particular inheritance:

First law—each factor segregates from each other into the GAMETES. It is now recognized that ALLELES are present on HOMOLOGOUS CHROMOSOMES, which separate during MEIOSIS.

Second law—factors for different traits undergo INDEPENDENT ASSORTMENT.

During gamete formation, the segregation of one gene

mesophyll the internal tissues of a leaf that are between the upper and lower epidermal layers. Mesophyll tissue contains CHLOROPLASTS, which are concentrated at a site that allows maximum absorption of light for PHOTOSYNTHESIS.

messenger RNA (mRNA) a single-stranded RNA molecule that contains ribose sugar, phosphate and the following four bases—ADENINE, CYTOSINE, GUANINE, and URACIL. Messenger RNA has the important role of copying genetic information from DNA in the nucleus and carrying this information in the form of a sequence of bases to RIBOSOMES in the cell, where TRANSLATION occurs to synthesize the specified protein that was encoded in the nuclear DNA.

metabolism the chemical and physical processes occurring in living organisms, which comprise two parts: CATABOLISM and ANABOLISM. ENZYMES control metabolic reactions, which tend to be similar throughout the plant and animal kingdoms.

metal substances that have a "metallic" lustre or sheen and are generally ductile, malleable, dense and good CONDUCTORS of electricity and heat. Elements with these properties are generally electropositive, i.e. give up electrons, becoming positively charged (e.g. Na^+) when combining with other RADICALS. When combining with water, BASES result (e.g. NaOH, sodium hydroxide), and their chlorides (e.g. NaCl, sodium chloride) are stable towards water. Not all elements normally considered metals show all these properties. Elements with characteristics of both metals and nonmetals are termed METALLOIDS.

metal fatigue *see* **fatigue of metals**.

metalloid an ELEMENT exhibiting some properties associated with metals and some associated with nonmetals. Metalloids also exhibit AMPHOTERISM.

metallurgy the scientific study of metals and their alloys, including extraction from their ORES, and processing for use.

metamorphic rock one of the three main ROCK types, formed by the alteration or recrystallization of existing rocks through the primary agents of temperature and/or pressure. There are four types, depending upon the pressure/temperature regime in operation: regional, which are formed by the high pressure and temperature accompanying orogenic (mountain-building) events; contact metamorphic rocks, which are generated by proximity to an IGNEOUS intrusion with high temperature but low pressure; dynamic, which are generated during faulting and thrusting (the sliding of rock masses against each other, producing intense pressure); and burial metamorphism, which is due to high pressures with low temperatures.

metamorphosis the period of change in form of an organism from the larval to the adult state.

metaphase a stage in MITOSIS or MEIOSIS in cells of EUCARYOTES. During metaphase, the chromosomes

microtubule

are organized and attached to the equator of the spindle by their CENTROMERES. Metaphase occurs only once in mitosis but twice in meiosis.

metastasis the process of malignant cancerous cells spreading from the affected tissue to create secondary areas of growth in other tissues of the body.

meteorite *see* **asteroid**.

meteorology the study of the processes and conditions (e.g. pressure, wind speed, temperature) in the earth's ATMOSPHERE. The resulting data enables predictions to be made as to likely future weather patterns.

methane the first member of the HOMOLOGOUS series of ALKANES. It is a colourless, odourless gas with the chemical formula CH_4. Methane is the main constituent of coal gas and is a byproduct of any decaying vegetable matter. It is flammable and is used in industry as a source of hydrogen.

mica a group of silicate minerals that occur characteristically in sheet form. They are complex hydrated alumino-silicates with numerous members depending upon the cations present (K, Na, Mg, Fe). Micas occur in all ROCK types.

micrometer any instrument used for the accurate measurement of minute objects, distances or angles.

micrometre the unit of length that equals 10^{-6} of a metre and has the symbol μm.

microtubule a long, hollow fibre of PROTEIN that is found in all higher plant and animal cells. Microtubules have different functions in different

microwave

cells, e.g. they form the spindle during MITOSIS, give strength and rigidity to the tentacles of some unicellular organisms, and they are also found in parts of nerve cells.

microwave the part of the electromagnetic spectrum (*see* ELECTROMAGNETIC WAVES) with a wavelength range of approximately 10^{-3} to 10m and a frequency range of 10^{11} to 10^{7}Hz. When absorbed, microwaves produce large amounts of heat, a useful property for the economical and quick cooking of food. Microwaves are easily deflected, and as they have a shorter wavelength range than radiowaves they are more suitable for use in RADAR systems as they can detect smaller objects.

Milky Way *see* **galaxy**.

mimicry an adaptively evolved resemblance of one species to another species. Mimicry occurs in both the animal and plant kingdoms but is predominantly found in insects. The main types of mimicry are:

(1) Batesian mimicry, named after the British naturalist H. W. Bates (1825-92)—where one harmless species mimics the appearance of another, usually poisonous, species. A good example of this is the nonpoisonous viceroy butterfly mimicking the orange and black colour of the poisonous monarch butterfly. The mimic benefits as, although harmless, any predator learns to avoid it as well as the poisonous species.

(2) Mullerian mimicry, named after the German zoologist J. F. T. Müller (1821-97)—where different

species, which are either poisonous or just distasteful to the predator, have evolved to resemble each other. This resemblance ensures that the predator avoids any similar-looking species.

mineralogy the study of any chemical element or compound extracted from the earth. Mineralogy examines the mode of formation and physico-chemical properties of minerals. These are generally solid or crystalline and can be classified according to their chemical constitution, i.e. molecular, metallic or ionic composition or crystallography. Another method of classifying minerals is according to their comparative hardness, and the Mohs' scale arranges them in relative order from the softest, talc (no. 1) to diamond (no. 10). Each mineral will scratch a mineral lower on the scale. The full list is: 1—talc; 2—gypsum; 3—calcite; 4—fluorite; 5—apatite; 6—orthoclase; 7—quartz; 8—topaz; 9—corundum; and 10—diamond.

mirage a visual phenomenon due to the REFLECTION and REFRACTION of light. A mirage is seen wherever there is calm air with varying temperatures near the earth's surface. A common mirage in the desert is caused by the refraction of a downward light ray from the sky, so that it seems to come from the sand and, to any onlooker, it would appear that the sky is reflected in a pool of water. As the inverted image of a distant tree is also usually formed, the overall effect resembles a tree being reflected in the surrounding water.

mitochondrion

mitochondrion (*plural* **mitochondria**) a double-membrane bound ORGANELLE found in EUCARYOTIC cells. Mitochondria contain circular DNA, RIBOSOMES and numerous ENZYMES, which are specific for essential biochemical processes. Thus, the mitochondrion is the site of such processes as haem synthesis (forms part of HAEMOGLOBIN) and the stages of AEROBIC RESPIRATION that generate most of a cell's ATP. Due to their ATP-producing function, mitochondria are especially abundant in very active, and hence high energy-requiring, cells such as muscle cells or the tails of human sperm cells.

mitosis the process by which a NUCLEUS divides to produce two identical daughter nuclei with the same number of CHROMOSOMES as the parent nucleus. Mitosis occurs in several phases:

(1) Prophase—the condensed chromosomes become visible, and it is apparent that each chromosome consists of two CHROMATIDS joined by a CENTROMERE.

(2) METAPHASE—the nuclear membrane disappears, a spindle forms, and each chromosome becomes attached by its centromere to the equator of the spindle fibres.

(3) Anaphase—each centromere splits, and one chromatid from each pair moves to opposite poles of the spindle.

(4) TELOPHASE—a nuclear membrane forms around each of the group of chromatids (now regarded as chromosomes), and the cytoplasm divides to produce

two daughter cells. The stage before and after mitosis is called interphase, and the chromosomes are invisible during this phase as they have decondensed. This is an extremely important part of a cell's cycle as DNA replicates and required proteins are synthesized during interphase.

Mobius strip a ribbon of paper where one end has been twisted through 180 degrees before joining to the other end. The result is a single continuous surface containing a continuous curve.

moderator a substance used in a nuclear reactor to slow down fast NEUTRONS generated by NUCLEAR FISSION. The substances contain a light element e.g. DEUTERIUM in heavy water, which absorbs some energy upon impact with the neutron, but avoids capturing the particle. These slower neutrons are then more likely to participate in the ongoing fission process.

modulus the measure of the MAGNITUDE of a quantity regardless of its sign. The modulus of a REAL NUMBER $|x|$ always gives a positive value, e.g. $|-6| = 6$. The modulus of a complex number $|a + ib|$ is the square root of the sum of the squares of the real part (a) and the imaginary part (ib, where $i = \sqrt{-1}$). Thus for the complex number, $|6 + i8|$, the modulus is as follows:

$$|6 + i8| = \sqrt{6^2 + 8^2} = \sqrt{36 + 64} = \sqrt{100} = 10$$

Mohs' scale *see* **mineralogy**.

molality the concentration of a solution expressed in MOLES of solute per one kilogram of solvent. Molality has symbol m and units of mol kg^{-1}.

molarity

molarity the number of MOLEs of a substance dissolved in one litre of solution. Molarity has symbol c and units mol litres^{-1}.

mole the amount of substance that contains the same number of ELEMENTARY PARTICLES as there are atoms in 12 grams of carbon. The number of particles in one mole is called AVOGADRO'S CONSTANT and equals 6.023×10^{23} mol^{-1}.

molecular formula the chemical formula that indicates both the number and type of any atom present in a molecular substance. For example, the molecular formula for the alcohol ETHANOL is C_2H_6O and indicates that the molecule consists of two carbon atoms, six hydrogen atoms and one oxygen atom.

molecularity the number of particles that are involved in a single step of a reaction mechanism. One step of a mechanism may be termed unimolecular, bimolecular or termolecular, depending on whether there are one, two or three reacting particles. The reacting particles can be ATOMS, IONS or MOLECULES.

molecular weight the total of the atomic weights of all the atoms present in a molecule.

molecule the smallest chemical unit of an element or compound that can exist independently. Any molecule consists of ATOMS bonded together in a fixed ratio, e.g. an oxygen molecule (O_2) has two bonded oxygen atoms and a carbon dioxide molecule (CO_2) has two oxygen atoms bonded to one carbon atom.

mole fraction the ratio of the number of MOLEs of a

momentum the property of an object that is directly proportional to both its MASS (m) and VELOCITY (v). Momentum (p) is a VECTOR quantity calculated using the equation p = mv (units kgms^{-1} or Ns^{-1}). Changes in momentum mainly occur as the result of an interaction between two bodies, and during any interaction the momentum of a body is always conserved, provided no external force such as FRICTION acts on it. If a body is in rotational motion around an axis, then it is said to have angular momentum. Angular momentum of an object is the product of its momentum and its perpendicular distance from the fixed axis.

monobasic acid an ACID that contains only one replaceable hydrogen atom per molecule. A monobasic acid will produce a normal SALT during a reaction with a suitable metal.

monocotyledon the subclass of flowering plants that have a single seed leaf (cotyledon). Any flowering plant that contains its seeds within an ovary (sometimes forming a fruit) belongs to one of two subclasses, either monocotyledon or dicotyledon (which has two seed leaves). Monocotyledons also differ from dicotyledons in that they have narrow, parallel-veined leaves, their vascular bundles are scattered throughout the stem, their root system is usually fibrous, and their flower parts are arranged in threes or multiples of threes. Most monocotyledons are small plants such as tulips, grasses or lilies.

monocyte a large phagocytic cell (*see* PHAGOCYTOSIS), capable of motion, which is present in blood. Monocytes originate in the bone marrow and, after a short residence in the blood, move into the tissues to become MACROPHAGES.

monomer a simple molecule that is the basic unit of POLYMERs. Most monomers contain carbon-carbon double bonds but just have single bonds after they have undergone ADDITION POLYMERIZATION to form the long-chained polymer. ETHENE is a type of monomer that reacts with other ethene molecules under high pressure and temperatures to form the polymer, polyethene (polythene).

monozygotic twins *see* **identical twins**.

mordant a chemical that is used in dyeing when the dye will not fix directly onto the fabric. The mordant (mainly weak basic HYDROXIDEs of aluminium, chromium and iron) impregnates the fabric, and the dye then reacts with the mordant, thus fixing it to the fabric.

morphine a crystalline ALKALOID occurring in opium. It is used widely as a pain reliever but misuse can be dangerous.

motility the ability to move independently.

mutation a change, whether natural or artifical, spontaneous or induced, in the constitution of the DNA in CHROMOSOMES. Mutation is one way in which genetic variation occurs (*see* NATURAL SELECTION) since any change in the GAMETEs may produce an inherited change in the characteristics of later generations of the organism. Mutations can be initiated by ionizing radiations and certain chemicals.

myelin sheath a fatty substance (myelin) that surrounds axons in the central nervous system of vertebrates and functions as an insulating layer.

N

nanometre a unit of measurement that is used for extremely small objects. One nanometre equals one thousand millionth of a metre (10^{-9}) and has the symbol nm.

natural gas the description usually applied to gas associated with PETROLEUM, which originated from organic matter. The gas is a mixture of HYDROCARBONS, mainly methane and ethane with propane. Also present in small amounts may be butane, nitrogen, carbon dioxide and sulphur compounds (*see also* FOSSIL FUELS).

natural logarithm *see* **logarithm**.

natural number any of the set of positive integers also known as the counting numbers: 1, 2, 3, 4 ....

natural selection the process by which evolutionary changes occur in organisms over a long period of time. Darwin explained natural selection (*see* DARWINISM) by arguing that organisms that are well adapted to their environment will survive to produce many offspring, whereas organisms that are not so well adapted to their enviromnent will not. As the better adapted organisms are successfully transmitting their genes from one generation to the next, these are the organisms that are "selected" to survive.

nautical mile the standard international unit of distance used in navigation. One nautical mile is defined as 1852 metres.

Nernst, Walther Hermann (1864-1941) a German physical chemist who was awarded the 1920 Nobel prize for chemistry for his proposal of the heat theorem, which was formulated as the third law of THERMODYNAMICS. In 1889, he also developed the Nernst equation, which is used to determine the electromotive force of a cell that contains non-standard constituents.

neuron another name for the nerve cell that is necessary for the transmission of information in the form of impulses along its body. The impulses are carried along long, thin structures of the neuron, known as the axon, and are received by shorter, more numerous structures called dendrites. It has been discovered that the transmission of impulses is faster in axons that are surrounded by MYELIN SHEATHS than those that are unmyelinated.

neutral a term indicating that a solution or substance is neither acidic or alkaline. The most well-known example of a neutral substance is pure water, which should have a pH of seven.

neutralization a reaction that either increases the pH of an acidic solution to neutral seven or decreases the pH of an alkaline solution to seven. Acid-base neutralizations occur in the presence of an INDICATOR, which undergoes a colour change when the reaction is complete.

neutrino *see* **elementary particle**.

neutron an uncharged particle that is found in the nucleus of an ATOM. The MASS of the neutron is 1.675×10^{-24}g, which is slightly larger than the mass of the PROTON.

neutron star a small body with a seemingly impossibly high density. A star that has exhausted its fuel supply collapses under gravitational forces so intense that its ELECTRONS and PROTONS are crushed together and form NEUTRONS. This produces a star 10 million times more dense than a WHITE DWARF—equivalent to a cupful of matter weighing many million tons on earth. Although no neutron stars have definitely been identified, it is thought that PULSARs may belong to this group.

newton the standard international unit of force. One newton (N) is the force that gives one kg an acceleration of one ms^{-2}.

Newton's laws of motion the fundamental laws of mechanics, which describe the effects of force on objects. Developed by Isaac Newton (1643-1727), the famous scientist, the three laws of motion state:

First law—any object will remain in a state of rest or constant linear motion provided no unbalanced force acts upon it.

Second law—the rate of change of momentum is proportional to the applied force and occurs in the linear direction in which the force acts.

Third law—every action has a reaction, which has a

force equal in magnitude but opposite in its direction. Until very recently, the above laws could not be directly demonstrated as they are idealized relationships that take place in the less idealized systems here on earth, where the effects of, say, FRICTION have to be considered.

niche all the environmental factors that affect an organism and its community. Such factors include spatial, dietary and physical conditions necessary for the survival and reproduction of a SPECIES. Only one species can occupy a specific niche within a community, as the coexistence of any species is subject to distinctions in their ecological niches.

nitric acid a colourless, corrosive acid liquid, HNO_3. It is a powerful oxidizing agent, attacking most metals and producing nitrogen dioxide (NO_2). It is prepared on a large scale by passing a mixture of AMMONIA (NH_3) and air over heated platinum, which acts as a CATALYST. It is used widely in the chemical industry.

nitrification the process by which soil bacteria convert ammonia (NH_3) present in decaying matter into nitrite (NO_2^-) and nitrate (NO_3^-) ions. The produced nitrate ions are taken up by plants and are used in their protein synthesis. Nitrification requires free oxygen, as the bacteria *Nitrosomonas* and *Nitrobacter* oxidize ammonia and nitrites respectively.

nitrogen cycle the regular circulation of nitrogen due to the activity of organisms. Nitrogen is found in all

nitrogen fixation

living organisms and forms about 80 per cent of the atmosphere (this proportion is maintained by the nitrogen cycle). The start of the nitrogen cycle can be regarded as the uptake of free nitrogen in the atmosphere by bacteria (NITROGEN FIXATION) and the uptake of nitrate (NO_3^-) ions by plants. The nitrogen is incorporated into plant tissue, which in turn is eaten by animals. The nitrogen is returned to the soil by the decomposition of dead plants and animals. NITRIFICATION converts the decomposing matter into nitrate ions suitable for uptake by plants.

nitrogen fixation the process by which free nitrogen (N_2) is extracted from the atmosphere by certain bacteria. Some free-living bacteria can use the nitrogen to form their AMINO ACIDS, while other nitrogen-fixing bacteria live in the root nodules of leguminous plants (peas and beans) and provide the plants with nitrogenous products. This enables the plant to survive in nitrogen-poor conditions while the bacteria has access to a carbohydrate supply in the plant (a symbiotic relationship). The nitrogen-fixing bacteria are able to convert free nitrogen into nitrogenous products because of the presence of the enzyme nitrogenase within their cells.

noble gases the elements comprising group 8 of the PERIODIC TABLE. Noble gases are usually referred to as INERT GASES because of their relative unreactivity.

node the site on a plant stem where the bud and leaves arise.

nonagon a nine-sided polygon that, if regular, has equal sides all with interior angles of 140°.

nondisjunction the failure of chromosomal pairs or sister CHROMATIDS to separate during MITOSIS or MEIOSIS respectively. Nondisjunction produces daughter cells containing an unequal number of CHROMOSOMES, either too many or too few.

normal a line perpendicular to the tangent of a curve or contact point of a line or plane.

normal salt any SALT formed by an ACID, which loses more than one hydrogen ion (H^+) per molecule during a NEUTRALIZATION reaction. It should be noted, however, that the production of a normal salt is dependent on the quantities of the acid and BASE used in the reaction.

Northern Lights *see* aurora.

nuclear fission a process that splits a heavy nucleus into two lighter nuclei. Nuclear fission produces more stable nuclei and emits huge amounts of energy. It is the process involved in the atom bomb, when uranium is bombarded with NEUTRONS, resulting in the splitting of uranium, which releases more neutrons, and a CHAIN REACTION ensues. Nuclear fission can be controlled by using synthetic rods to absorb excess neutrons. This is the basis of NUCLEAR POWER, which generates the energy used to propel nuclear submarines and to produce electricity from nuclear power plants.

nuclear fusion a process that fuses two lighter nuclei

into a heavier, more stable nucleus. Nuclear fusion releases tremendous amounts of energy and is believed to be the energy source of the sun as hydrogen nuclei are converted into helium nuclei. Very high temperatures are required for nuclear fusion, and the hydrogen bomb, which is based on fusion, needs the large energy source of NUCLEAR FISSION to initiate the explosion. Controlled nuclear fusion would be a greater energy source than controlled nuclear fission and would not generate radioactive byproducts, but, unfortunately, the byproduct of nuclear fusion—a high-temperature, dense gas—cannot, at present, be confined.

nuclear membrane a double membrane that surrounds the nucleus of EUCARYOTIC cells. There is a space between the outer and inner nuclear membranes, but both membranes seem to fuse together at regions containing nuclear pores. Nuclear pores contain proteins that probably control the exchange of material between the cell nucleus and its CYTOPLASM.

nuclear power the production of energy from the controlled NUCLEAR FISSION that involves uranium and plutonium as fuels. Nuclear power is used to generate electricity by removing the huge amount of energy released during fuel fission away from the core reactor to the outside, where it is converted to steam and generates electricity by driving turbines.

nucleic acid a linear MOLECULE that acts as the genetic information store of all cells. Nucleic acids occur in two forms, deoxyribonucleic acid (DNA) and ribonucleic acid (RNA), but both forms are composed of four different NUCLEOTIDEs, which react to form the long chain-like molecule. DNA is found inside the nucleus of all EUCARYOTEs as it is the major part of CHROMOSOMEs, but RNA is found outside the nucleus and is essential for TRANSCRIPTION and TRANSLATION during protein synthesis.

nucleolus a membrane-bound object found within the NUCLEUS. The nucleolus contains gene sequences that code for ribosomal RNA, ribosomal RNA itself, and proteins necessary for rRNA synthesis.

nucleophile a reactive molecule that will readily "donate" its unshared pair of electrons. Nucleophiles will attack the low electron density regions of other molecules.

nucleotide a MOLECULE that acts as the basic building block of the NUCLEIC ACIDS, DNA and RNA. The structure of a nucleotide can be divided into three parts—a five-carbon sugar molecule; a phosphate group; and an organic base. The organic base can be either a PURINE, e.g. ADENINE, or a PYRIMIDINE, e.g cytosine.

nucleus (*plural* **nuclei**) in biology, the ORGANELLE that contains the chromosomes of EUCARYOTic cells. Molecules enter and leave the nucleus via the pores in the NUCLEAR MEMBRANE. Such molecules include AMINO ACIDs and MESSENGER RNA. In chemistry and physics, the term nucleus refers to the small, positively charged core of an ATOM that contains the PRO-

null hypothesis

TONS and NEUTRONS. The electrons of an atom orbit the nucleus.

null hypothesis a hypothesis that examines the existence of a specific relationship by enabling it to be statistically tested. A null hypothesis assumes that the expected results from an experiment have just arisen from chance with no significant change occurring. These expected results are statistically compared with the observed results derived from the experiment. Any significant difference between the expected and the observed results will be taken as evidence to support the experimental hypothesis and to rule out the null hypothesis.

numerator the number or quantity to be divided by the denominator of a fraction. For example, the fraction $^3/_4$ has a numerator of 3.

O

oblique angle any angle that does not equal 90° (right angle) or any multiple of 90°.

obtuse angle any angle that lies between but does not equal 90° and 180°.

octahedron a geometrical solid that consists of eight planes, each bound by their edges. If the eight faces are all equilateral triangles, then the octahedron is said to be regular.

oestrogens a group of female sex hormones, including some sterols, e.g. oestradiol, which is one component of oral contraceptives.

ohm (Ω) the unit of electrical resistance. Between two points of an electrical CONDUCTOR, one VOLT (V) is needed to force a current (I) of one AMPERE through a RESISTANCE (R) of one ohm, i.e. $V = IR$. This is known as **Ohm's law** (after the German physicist who formulated it, Georg Simon Ohm [1787-1854]), which can be rewritten: $R = V/I$.

oil a greasy liquid sustance obtained from animal or vegetable matter (*see* FATS) or from mineral matter (*see* PETROLEUM).

oil shale a dark, fine-grained shale (*see* SEDIMENTARY ROCK) containing organic substances that produce liquid HYDROCARBONS on heating, but do not contain free PETROLEUM.

omnivore any organism that eats both plant and animal tissue.

oncogene any gene directly involved in cancer. Oncogenes may be part of a specific VIRUS that has managed to penetrate and replicate within the host's cell, or they may be part of the individual's GENOME, which has been transformed by radiation or a chemical.

oocyte a cell that undergoes MEIOSIS to form the female reproductive cell (egg or ovum) of an organism. In humans, a newborn female already has primary oocytes, which will undergo further development when puberty is reached but which will only complete secondary meiosis to form the secondary HAPLOID oocyte if fertilization occurs.

open chain a compound with an open chain not a ring structure, as in aliphatic compounds, e.g. ALKANES, ALKENES and ALKYNES and their derivatives.

optical activity (*or* **optical rotation**) is when a solution of a substance rotates the plane of transmitted polarized light (*see* POLARIZATION). It occurs with optical ISOMERs, with one form rotating the light in one direction and the other rotating the light by the same amount in the opposite direction (DEXTROROTATORY and LAEVOROTATORY compounds). A racemic mixture is optically inactive because it contains equal amounts of both forms.

optical fibre a small, thin strand of pure glass that uses internal reflection to transmit light signals. Optical fibres are more efficient than conventional cables as they are much smaller, thus requiring less space, and have a higher data-carrying capacity.

optical isomerism the existence of two chemical compounds that are ISOMERs, which form non-superimposable mirror images.

orbitals orbitals are a means of expressing, rationalizing and correlating atomic structure, bonding and similar phenomena. The BOHR THEORY postulated the positioning of ELECTRONS in definite orbits about a central NUCLEUS. However, it was soon discovered that this was too simple and that electrons behave in some ways as waves, which makes their spatial position more imprecise. Hence the old "particle in an orbit" picture was replaced by an electron "smeared out" into a charge cloud or orbital, which represents the probability distribution of the electron. An atomic orbital is thus one associated with an atomic nucleus and has a shape determined by QUANTUM NUMBERS. Various types of orbital, designated s, p, d, etc, are distinguished. An s orbital is spherical, and a p orbital is dumbbell-shaped. When two atoms form a COVALENT BOND, a molecular orbital with two electrons is formed, associated with both nuclei (*see* σ and π BOND). The overlapping of atomic orbitals in a carbon-carbon single bond (e.g. ETHANE) creates a molecular orbital centred on the line joining the two nuclei. In a carbon-carbon double bond (e.g. ETHENE), the second of the two bonds is created by two overlapping p orbitals, forming the π bond. The two overlapping dumbbells create two sausage-like spaces of electron "cloud" on each side of the line joining the nuclei. BENZENE has torus-shaped (doughnut-shaped) molecular orbitals on each side of the ring due to overlap and merging of p atomic orbitals.

order of magnitude the approximate size of an object or quantity usually expressed in powers of 10.

ordinal scale a statistical scale that arranges the data in order of rank in the absence of a numerical scale with regular intervals. Ordinal scales are ideal for data that contain relationships such as bad, good, better, best, as the data can be put in rank order but no regular interval can be measured between the ranked judgements.

ordinate the vertical or y-axis in a geometrical diagram for CARTESIAN CO-ORDINATES. For example, a point with co-ordinates (2,-6) has an ordinate of -6.

ore any naturally occurring substance that contains commercially useful metals or other compounds. The extraction of the desired metal will only proceed if the process is both economically and chemically feasible. Some relatively unreactive metals such as copper and gold exist as native ores with no need of extraction, but most metals are obtained by extracting them from their oxygen-containing (oxide) ores.

organelle any functional entity that is bound by a membrane to separate it from the other cell constituents. Organelles are found in the cells of all EUCARYOTES and include the CHLOROPLASTS and vacuoles of plant cells in addition to the NUCLEUS, MITOCHONDRIA, GOLGI APPARATUS, ENDOPLASMIC RETICULUM, and other small vesicles of both animal and plant cells.

organic chemistry the branch of chemistry concerned with the study of carbon compounds. Organic chemistry studies the typical bond arrangements and properties of carbon compounds containing hydrogen and, less frequently, oxygen and nitrogen. As most organic compounds are derived from living organisms, two major areas for study are the biologically important organic compounds and the commercially important organic compounds, e.g. ALKANES derived from oil.

origin the point of intersection of the horizontal (x-axis) and the vertical (y-axis) axes in a two-dimensional diagram for CARTESIAN CO-ORDINATES. Thus the origin has co-ordinates of (0,0).

oscillation the regular production of a fluctuating position or state. Oscillation can mean the cycle, vibration or rotation of the object in question. In the case of a simple pendulum, oscillation refers to its regular swinging motion and, when used in connection with electrical circuits, oscillation refers to the production of an ALTERNATING CURRENT.

oscillator a CIRCUIT or device that produces an ALTERNATING CURRENT or voltage of a specific FREQUENCY as its output signal. Two necessary components of an

osmosis

oscillator are the capacitor, which produces the electric field, and the inductor (coil), which produces the magnetic field (*see* ELECTROMAGNETIC INDUCTION). The alternating current of an oscillator is fundamentally caused by energy transfers between the electrical energy stored in the capacitor and the magnetic energy stored in the inductor. The capacitor also determines the frequency, as its value can be changed to get the desired frequency.

osmosis the process in which solvent molecules (usually water) move through a semi-permeable membrane to the more concentrated solution. Many mechanisms have evolved to prevent the death of animal cells either by too much water entering a cell by osmosis, causing it to rupture, or by too much leaving by osmosis, causing it to shrink (plasmolysis). Such mechanisms include the presence of a pump within the membrane of animal cells, which actively regulates the concentration of vital cellular IONs and the excretion of salt through the gills of marine bony fish to remove the salt gained by diffusion and drinking.

oviparous a term describing animal reproduction where the development of the embryo and subsequent hatching occurs outside the female's body. Oviparous reproduction is found in birds, most fish, and reptiles.

ovoviviparous the term to describe the development of offspring within the body of the female in the absence of a placenta. Ovoviviparous reproduction is found in certain species of fish, reptiles and insects,

oxidation any chemical reaction that is characterized by the gain of OXYGEN or the loss of ELECTRONs from the reactant. Oxidation can occur in the absence of oxygen, as a molecule is also said to be oxidized if it loses a hydrogen atom.

oxide a compound formed by the combination of OXYGEN with other elements, with the exception of the INERT GASes.

oxidizing agent any substance that will gain ELECTRONs during a chemical reaction. Oxidizing agents will readily cause the OXIDATION of other atoms, molecules or compounds, depending on the strength of the oxidizing agent and the reactivity of the other substance. The following are all examples of oxidizing agents arranged in order of increasing oxidizing strength—sodium ions (Na^+), sulphate ions (SO_4^{2-}), and oxygen molecules (O_2).

oxygen a colourless and odourless gas, which is essential for the respiration of most life forms. It is the most abundant of all the elements, forming 20 per cent by volume of the atmosphere; about 90 per cent by weight of water; and 50 per cent by weight of ROCKs in the crust. It is manufactured by the FRACTIONAL DISTILLATION of liquid air and is used for welding, anaesthesia and rocket fuels.

ozone a denser form of oxygen that exists as three

ozone layer

atoms per molecule (O_3). Ozone is a more reactive gas than the more common diatomic molecule (O_2), and can react with some hydrocarbons in the presence of sunlight to produce toxic substances that are irritants to the eyes, skin and lungs. Minute quantities of O_3 are found in sea water. It forms the earth's OZONE LAYER, 15 to 30 kilometres above the earth's surface.

ozone layer a region of the earth's atmosphere containing ozone that acts as a barrier against the ULTRAVIOLET RADIATION from the sun. Scientists and environmentalists have announced that large holes are appearing in the ozone layer as a result of the widespread use of the ozone-depleting chemicals called chlorofluorocarbons (CFCs). CFCs are used in many industrial processes and, because of their unreactive qualities, as aerosol propellants. In the earth's atmosphere, however, they will readily react with, and destroy, ozone in the presence of sunlight. The depletion of the ozone layer is cause for concern as increased exposure to ultraviolet radiation will increase the incidence of skin cancers and eye cataracts. To prevent further damage to the ozone layer, industrial nations are being called upon to greatly reduce the use of CFCs.

P

palaeontology the scientific study of FOSSILS.

Pangaea *see* **continental drift**.

parabola a plane curve traced out by a point moving so that its distance from a fixed point (focus) is equal to its perpendicular distance from a fixed straight line (directrix).

parabolic rule *see* **Simpson's rule**.

parallel circuit a CIRCUIT in which each component has the same POTENTIAL DIFFERENCE but has a different amount of current flowing through it. The amount of current flowing through each component depends upon a phenomenon called RESISTANCE, and in parallel circuits the total resistance (R) of any circuit components is given by the following relationship:

$$1/R = 1/R1 + 1/R2 + 1/R3 \ldots \text{etc.}$$

parallelogram a four-sided POLYGON, which has opposite sides that are parallel and equal in length. Parallelograms can have four sides all of equal length (equilateral parallelogram, i.e. a rhombus), four equal angles (equiangular parallelogram, i.e. a rectangle) or have all four sides and angles equal, i.e. a square.

parameter an arbitrary constant or variable that determines the specific form of a mathematical equation,

parametric equation

as a and b in $y = (x - a)^2 + b$. Changing the value of the parameter generates various cases of the phenomena represented.

parametric equation an equation where the co-ordinates of a point appear dependent on parameters. Any point (x, y) on a curve or on a surface may be expressed as functions of a third variable, t, such that x and y are functions of t; $x = f(t)$, $y = g(t)$.

parasite any organism that obtains its nutrients by living in or on the body of another organism (its host). Parasites can be either completely dependent on their host for survival (obligate) or are able to survive without their host (facultative). The extent of the damage on the health of the host by parasitic infestation can range from being virtually harmless to so severe that it causes the death of the host. Highly evolved parasites are so well adapted, however, that the host is able to survive and reproduce as normal, thus providing the parasite with a permanent supply of nutrients.

parenthesis (*plural* **parentheses**) the curved brackets () used to group terms or as a sign of aggregation in a mathematical or logical expression.

parthenogenesis the development of a new individual from an unfertilized egg. Parthenogenesis is most common among the lower invertebrates, such as insects and flatworms. For example, the process can be part of the honey bee life cycle if the HAPLOID eggs laid by the queen remain unfertilized by sperm. The larvae from these eggs will develop into the male bees (drones), whose only function is to produce sperm. If the eggs laid by the queen are fertilized then the larvae develop into sterile female worker bees or fertile queens, depending on the food supply. Parthenogenesis also occurs in some plants, such as the common dandelion, and it can be induced artificially in many species by stimulation of the egg cell.

partial fractions the simple FRACTIONs into which a larger fraction may be separated so that the sum of the simpler fractions equals the original fraction.

pascal the unit of PRESSURE named after the French mathematician, philosopher and physicist, Blaise Pascal (1623-1662). One pascal (Pa) is defined as the FORCE of one NEWTON acting on a square metre, i.e. 1 Pa = 1 Nm^{-2}. One atmosphere pressure (760mm Hg, the air pressure at sea level) is approximately 100 kilopascals (kPa).

Pascal's law of fluid pressures the PRESSURE of a fluid is the same at any point since any applied pressure will be transmitted equally to all points of the containing vessel. Pascal discovered this principle while mountaineering with his father. He realized that the column of mercury in the barometers he carried would vary in length—essentially the principle behind all hydraulic systems.

Pascal's triangle the diagrammatical array of integers starting with one such that each number is the sum of the two numbers in the row directly above it. The result is a triangle of potentially infinite size, the beginning of which is as follows:

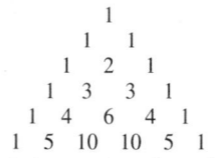

```
            1
          1   1
        1   2   1
      1   3   3   1
    1   4   6   4   1
  1   5  10  10   5   1
```

Pascal's triangle is an extremely useful method for determining the COEFFICIENTS when using the BINOMIAL THEOREM for expanding equations of $(a + b)n$, where the nth line of the triangle corresponds to n.

passive immunity the ability of an individual to resist disease using ANTIBODIES that have been donated by another individual rather than by producing its own antibodies. Passive immunity is obtained by young mammals from their mother's milk during the first few weeks of life as the newly born are virtually incapable of antibody production. In humans, breast-fed infants will receive most of their maternal antibodies from their mother's milk, but one antibody, IgG, will be found in all infants, whether breast-fed or bottle-fed, as IgG can cross the placenta during foetal development.

Pasteur, Louis (1822-1895) French chemist and bacteriologist who was the first to demonstrate that a colony could be grown in a culture medium that had been infected with a few cells of the micro-organism. This experiment showed that living cells had an inheritance of their own and helped discredit the theory of SPONTANEOUS GENERATION of life. Although Pasteur had been aware of the role of micro-organisms in FERMENTATION since 1858, he did not accept their role in causing disease until several years later. He demonstrated that attenuated forms of micro-organisms could be used in innoculation, providing immunization for the host, and in 1885 he produced the first rabies vaccine.

pasteurization a process developed by PASTEUR of partially sterilizing food by heating it to a certain temperature. Food is pasteurized before distribution as the process can destroy potentially harmful bacteria, e.g. heating milk for 30 minutes at 62°C destroys the bacteria responsible for tuberculosis, and increases the shelf life of food by delaying its FERMENTATION.

pathogen any organism that causes disease in another organism. Most pathogens that affect humans and other animals are bacteria or viruses, but in plants there is also a wide range of fungi that act as pathogens.

Pauling, Linus Carl (1901-) an American biochemist who determined the structure of crystals of simple molecules and pure proteins by using X-ray crystallography. With fellow colleagues, Pauling discovered one of the regular structures common to all proteins, the α-helix and, using ELECTROPHORESIS, isolated the abnormal HAEMOGLOBIN that causes an hereditary form of anaemia. He also devised the Pauling scale, which is a useful method of making qualitative com-

pentagon

parisons between the ELECTRONEGATIVITIES of the elements.

pentagon any plane shape that has five sides. A regular pentagon has sides of equal length and five interior angles each measuring 108°.

pentahedron any three-dimensional figure that has five plane faces.

peptide bond the chemical linkage formed when two AMINO ACIDS join together. As all amino acids have a common molecular structure, the reaction always involves the elimination of a water molecule as the amino group (NH_2) of one amino acid molecule joins to the carboxyl group (COOH) of another molecule.

perigee *see* **apogee**.

perimeter the total distance round the outside of a closed plane figure, such as the circumference of a circle.

period the time taken for a body to complete one full OSCILLATION, which can involve vibrational, rotational or harmonic motion. Period (T) has seconds (s) as its units, and it is the reciprocal of FREQUENCY, i.e. $T = 1/f$.

periodic function a mathematical function (e.g. sine or cosine) whose possible values all recur at regular intervals. The graph of the function $y = \sin x$, where x is the number of degrees, produces a curve that repeats itself every 360°, i.e. it has a period of 360°. In general, a function f(x) of a real or complex variable is periodic, with period T if $f(x + T) = f(x)$ for every value of x.

periodic table an ordered arrangement of the elements by their ATOMIC NUMBER. The elements are arranged by periods (horizontally, *see* APPENDIX 1), which correspond to the filling of successive shells, and by groups (vertically), which reflect the number of VALENCY ELECTRONS, i.e. the number in the outer shell.

peristalsis the involuntary muscular contractions responsible for moving the contents of tubular organs in one direction. Peristalsis occurs in the alimentary canal of animals as the alternate waves of contraction and relaxation of smooth muscle move food and waste products along.

permutation an ordered arrangement of a set of objects into specified groups. The number and order of component objects is important, e.g. the arrangement of four letters, ABCD, taken two at a time yields 12 permutations: AB, AC, AD, BC, BD, CD, BA, CA, DA, CB, DB, DC. The formula for the number of permutations that can be made from n dissimilar objects taken r at a time is $n!/(n-r)!$, where ! stands for FACTORIAL.

perpendicular any line or plane that meets another line or plane at a right angle (90°). If the perpendicular is formed by a line meeting a plane or the tangent to a curve, then the line is referred to as the NORMAL of that plane or curve.

petroleum (crude oil) a mixture of naturally occurring HYDROCARBONS formed by the decay of organic matter under pressure and elevated temperatures. Oil thus formed migrates from its source to a permeable reservoir rock, which is capped or sealed by an impermeable cover. The composition of the petroleum varies with the source and is separated initially by FRACTIONAL DISTILLATION into its major components (gas, liquids, wax, and residues such as bitumen). The liquids include petrol, paraffin oil, and other hydrocarbon liquids. CRACKING is used to break down some substances to create smaller molecules that can be used more readily. In addition to the production of various fuels, petroleum is the basis of the vast petrochemicals industry. *See also* FOSSIL FUELS.

pH the measure of concentration of hydrogen IONS (H^+) in an aqueous solution. The pH is the negative LOGARITHM (base 10) of H^+ ion concentration, calculated using the following formula:

$$pH = \log_{10}(1/(H^+))$$

The scale of pH ranges from 1.0 (highly acidic), with decreasing acidity until pH 7.0 (NEUTRAL) and then increasing alkalinity to 14 (highly alkaline). As the pH measurement is logarithmic, one unit of pH change is equivalent to a tenfold change in the concentration of H^+ ions.

phagocytosis the process by which cells bind and ingest large particles from the surrounding environment. In phagocytosis, the target particle binds to the cell's surface and is then completely engulfed by a bud formed by the plasma membrane of the cell. This process is used by simple unicellular organisms to ingest food particles and by certain LEUCOCYTES to engulf and destroy bacteria and old, broken cells.

phenol (C_6H_5OH—carbolic acid) as a solution in water, it is corrosive and poisonous. It is used as a disinfectant (with the typical "carbolic" smell) and in the manufacture of dyes and PLASTICS.

phenotype the detectable characteristics of an organism that are determined by the interaction between its GENOTYPE and the environment in which the organism develops. Organisms with identical genotypes may have different phenotypes, due to development in environments that differ in, for example, the availability of important nutrients or specific stimuli. It is unlikely, however, that organisms that have identical detectable phenotypes will have different genotypes unless they are HETEROZYGOTES. The expression of the dominant gene masks the presence of a recessive gene, as only the expressed gene affects the organism's phenotype.

pheromone a molecule that functions as a chemical communication signal between individuals of the same species. Pheromones are used extensively throughout the animal kingdom and have a wide range of functions. They can act as sexual attractants (very common in insects) and can help establish territories, as demonstrated by the frequent urination by dogs. Although pheromones are much rarer in plants than animals, one of the most economically and environmentally important pheromones is produced by a plant called the "Scary Hairy Wild Potato" (*Solanum berthaultii*). The leaves of this plant produce a pheromone that is identical to the warning signal produced by aphids. Breeding this aphid-repellent character into cultivated crops will reduce the financial loss from crop damage and reduce pollution as insecticides are needed less.

phospholipids lipids containing phosphoric acid (H_3PO_4) groups and nitrogenous bases. Phospholipids are found in brain tissue and egg yolks.

phosphorescence LUMINESCENCE that continues after the initial cause of excitation. The substance usually emits light of a particular WAVELENGTH after absorbing ELECTROMAGNETIC radiation of a shorter wavelength.

photon a QUANTUM of energy that is an intrinsic component of all ELECTROMAGNETIC WAVES. Photons are used to explain the quantum theory of light, where the properties of light are explained in terms of particles (photons), as opposed to the wave theory of light, where its properties are explained by the propagation of a wave and how it disturbs a medium. The energy of a photon is proportional to the FREQUENCY of the light beam.

photosynthesis the process by which plants make carbohydrates, using water, carbon dioxide (CO_2) and light energy, while releasing oxygen. Photosynthesis occurs in two stages, known as the CALVIN CYCLE and the LIGHT REACTIONS of photosynthesis. For photosynthesis to occur, an organism must contain light-trapping pigments, which capture light energy in the form of PHOTONS and use the photons to initiate a series of energy-transfer reactions. Some blue-green algae (cyanobacteria) and the CHLOROPLASTS of all plants contain the essential light-trapping pigment called CHLOROPHYLL that makes them capable of photosynthesis. Photosynthesis is an essential process for regulating the atmosphere as it increases the oxygen concentration while reducing the CO_2 concentration.

phototropism a growth movement exhibited by parts of plants in response to the stimulus of light. Plant shoots display positive phototropism as they grow towards the light source, but the roots tend to display negative phototropism as they grow away from the light source (*see* GEOTROPISM). Phototropism is caused by the unequal distribution of auxin (a plant growth hormone) as this substance has a higher concentration in the darker side of the plant and thus increases growth on this side by inducing cell elongation.

physics the study of matter and energy and changes in energy without chemical alteration. Physics includes the topics of magnetism, electricity, mechanics, heat, light and sound. The study of modern physics also encompasses quantum theory, atomic and nuclear physics. In combination with other disciplines, physics forms new topics, e.g. geophysics, biophysics.

pi (π) bond the COVALENT BOND formed when two atoms join to form a diatomic molecule. Pi bonds are discussed in terms of molecular ORBITALS, in which the shared electrons orbit the whole molecule rather than an atom. Pi bonds hold the molecule together by forming two regions of electron density above and below an axis between the bonded nuclei of the two atoms.

piezoelectric effect an effect of certain ANISOTROPIC crystals whereby opposite charges are generated on opposite crystal faces by the application of pressure. QUARTZ is such a crystal. One use of this phenomenon is in the crystal microphone.

pituitary gland *see* **endocrine system**.

place value notation when a number has more than one digit, the position, or place value, of each digit in the number is used to indicate what it is worth. For example, using the decimal system, the 6 in 362 and in 3620 stands for something different: in 362 it means 6 "tens" and in 3620 it means 6 "hundreds."

Planck's constant the proportionality constant (h) used in the equation to define the energy of a QUANTUM. Planck's constant has a value of 6.6262×10^{-34} Js and is named after the German mathematician and physicist Max Planck (1858-1947), who proposed the theory that radiant energy consisted of quanta.

planetoid *see* **asteroid**.

plankton very small organisms, of plant and animal origins, that drift in water. The plants (or phytoplankton) are mainly diatoms (unicellular ALGAE) which PHOTOSYNTHESIZE and form the basis of the food chains. The animals or zooplankton feed on the diatoms and include small crustaceans and the larval stages of larger organisms.

plasma in biology, the same as BLOOD PLASMA. In physics, essentially a high temperature gas of charged particles (ELECTRONs and IONs) rather than neutral atoms or molecules. A plasma is electrically neutral overall, but the presence of charged particles means that it can support an electric CURRENT. It is of significance to the study of controlled NUCLEAR FUSION.

plasmolysis *see* **osmosis**.

plastics a group name for mainly synthetic organic compounds, which are mostly POLYMERs (formed by polymerization) that, when subjected to heat and pressure, become plastic and can be moulded. There are two types: thermosetting and thermoplastic materials. Thermosetting plastics are materials that lose their plasticity after being subjected to heat and/or pressure. Thermoplastic materials become plastic when heated and can be heated repeatedly without changing their properties.

platelet *see* **blood**.

plate tectonics *see* **continental drift**.

platinum metals a block of six TRANSITION ELEMENTS with similar properties—ruthenium (Ru), rhodium (Rh), palladium (Pd), osmium (Os), iridium (Ir) and platinum (Pt). The platinum metals are commonly found together, with gold and silver. Of the group, platinum is of greatest importance. It is very stable and is used mainly for jewellery, special scientific equipment, and chemical electrodes. It is also used as a CATALYST (*see also* ZEOLITES).

point of inflection a point where a plane curve changes from the concave to the convex, relative to some fixed line, i.e. the point where it "crosses its tangent." At this point, the second DERIVATIVE of the function determining the curve is zero, i.e. $d^2y/dx^2 = 0$.

Poiseuille's formula an expression that examines the relationship between the volume flow rate (Q) of a pipe and the radius of the pipe (a), the pressure gradient along the pipe (p/l), and the viscosity of the liquid carried by the pipe (η). Poiseuille's formula for the average volume per second (m3s-1) is as follows:

$$Q = \frac{\pi p a^4}{8\eta l}$$

Poiseuille's formula does not apply to turbulent motion in pipes, but it is useful for examining pipes in which flow patterns have developed and can even be used to investigate blood flow in arteries or veins or water flow in plant stems or roots.

polar angle the angle between the positive (polar) axis and the radius vector in POLAR CO-ORDINATES.

polar co-ordinates the position of a point in space as represented by the co-ordinates (r, q), where q is the angle between the positive x-axis and a line from the origin to the point, and r the length of that line.

polar covalent bond the joining of two ATOMS due to the strong but unequal sharing of their ELECTRONS, which gives the bond, and thus the molecule formed, partial charges. Polar covalent bonds are formed between atoms that differ in their ability to attract electrons, with one atom having a greater ELECTRONEGATIVITY than another. For example, a hydrogen chloride molecule (HCl) has a polar covalent bond, as the chlorine atom is more electronegative than the hydrogen atom. The net result is that the chlorine end of the bond has a denser electron cloud because of its greater attraction for electrons. This causes the chlorine end of the molecule to have a partial negative charge, and as the whole HCl molecule is neutral, the hydrogen end has a corresponding partial positive charge.

polarization the process by which the particles of a light wave are made to vibrate in one particular plane rather than the many directions taken by particles of normal light. Only transverse waves, and not longitudinal waves, can be polarized, so all electromagnetic waves can be polarized but longitudinal waves, such as sound, cannot. Some natural crystals, e.g. quartz and calcite, can polarize light because of their internal structure, and polarization has many scientific uses.

polygon a closed plane figure with three or more straight line sides. Common polygons are figures such as the triangle, quadrilateral and pentagon. A square is an example of a regular polygon, one in which all sides and all angles are equal. A general equation exists for the sum of the interior angles of a polygon with n sides: sum of the interior angles = $180°$ (n - 2).

polyhedron a solid figure composed of four or more polygonal plane faces. The more faces it has, the closer it is to a sphere. A cube is an example of a regular polyhedron as all the faces and all the angles of a cube are equal.

polymer a large, usually linear MOLECULE that is formed from many simple molecules, MONOMERS. Natural polymers include starch, cellulose (found in the cell walls of plants), and PROTEINS. Many synthesized polymers, such as nylon and polythene, are formed from ALKENES by ADDITION POLYMERIZATION.

polynomial in mathematics, an algebraic expression consisting of three or more terms, each of which is the product of a constant and one or more variables raised to a positive or zero integrated power. In biology, polynomial denotes a species name of more than two terms.

polypeptide a single, linear MOLECULE that is formed from many AMINO ACIDS joined by PEPTIDE BONDS. Polypeptides differ greatly in the number of amino acids they contain (usually from 30 to 1000). Although there are only 20 different amino acids, there are a huge number of possible arrangements in a polypeptide or PROTEIN, as the amino acids can be in any order. Most proteins consist of more than one polypeptide rather than a single polypeptide chain.

porphyrins naturally occurring pigments that include CHLOROPHYLL and the haem part of HAEMOGLOBIN.

positron a particle with the same mass as the ELECTRON but a positive electrical charge. Positrons are produced during decay processes (*see* BETA DECAY) and themselves are annihilated on passing through matter (*see* ANTIMATTER).

potential difference the work done in driving a unit of electric charge (one COULOMB) from one point to another in a current-carrying CIRCUIT. The unit of potential difference is the VOLT, and the potential difference is frequently referred to as the VOLTAGE.

potential energy any energy stored within a body that can be used to do work. A body is said to have potential energy (U) when it has been raised from a resting point A against gravitation to resting point B. The potential energy of such a body can be derived from $U = mgh$, where m is the mass of the body, g is 9.8 ms^{-2} (GRAVITY), and h is the distance moved. If the body is released from resting point B, its potential energy is transformed into the energy of motion, KINETIC ENERGY. Potential energy is present in a spring that has been stretched and can also be stored in the form of chemical or electrical energy.

power the rate at which work is done by or against a FORCE. Power is also regarded as the rate at which energy is converted. The unit of power is the WATT (W), which is equal to the transfer of one energy joule per second, i.e. $1W = 1Js^{-1}$. Electrical power can be

power notation the use of a small number (an EXPONENT) placed next to an ordinary number to show how many times the ordinary number is multiplied by itself e.g.:

3^5 ("3 to the power 5") means $3 \times 3 \times 3 \times 3 \times 3$

power series a series of functions of the form:

(A) $a_0 + a_1x + a_2x_2 + ... + a_nx_n + ...$

or

(B) $a_0 + a_1(x - a) + a_2(x - a)^2 + ... + a_n(x - a)^n + ...$

where x is a real VARIABLE and a represents constant COEFFICIENTS.

precipitation the formation of an insoluble substance (precipitate) during a reaction between two solutions. Precipitation occurs because the IONs of the two substances involved exchange partners. For example, if a solution of silver nitrate, $AgNO_3(aq)$, is mixed with a solution of sodium chloride, $NaCl(aq)$ the insoluble silver chloride $AgCl(s)$ forms a precipitate. From the chemical equation for this reaction:

$AgNO_3(aq) + NaCl(aq) \longrightarrow AgCl(s) + NaNO3(aq)$

it can be seen that the chloride ion (Cl^-) has displaced nitrate.

pressure the FORCE exerted per unit area of a surface. The pressure of a gas is equal to the force that its molecules exert on the walls of the containing vessel, divided by the surface area of the vessel. The pressure of a gas will vary with its temperature and volume, as stated by BOYLE'S LAW, CHARLES' LAW, and the GAS LAWS. At any depth, the pressure in a liquid or in air equals the weight above the unit area, and therefore as the depth increases, the pressure also increases. This is also the reason for air pressure decreasing as height above sea level increases. The unit of pressure is the PASCAL, although air pressure is commonly measured using mercury BAROMETERs and hence has units of millimetres of mercury (mm Hg) corresponding to the varying mercury levels as air pressure changes.

pressure law *see* gas laws.

prime number a number that can only be divided by itself and 1. The first ten prime numbers are 2, 3, 7, 11, 13, 17, 19, 23, 29.

prism in mathematics, a solid with equal and parallel POLYGONS as ends and parallelograms as sides. In physics, a prism is triangular in shape and made of transparent material. They are used to deviate or disperse a ray in optical instruments or laboratory experiments.

procaryote any organism that lacks a true-membrane NUCLEUS and is either a bacterium or a blue-green algae (cyanobacteria). Procaryotes have a single CHROMOSOME and do not undergo MEIOSIS or MITOSIS as they lack the MICROTUBULES to form the spindle. Procaryotes replicate by a form of asexual reproduction, called binary fission, in which the two sister chromosomes are attached to separate regions on the cell membrane, which starts to fold to form a cleavage.

product rule

The cell eventually forms two daughter cells after the CYTOPLASM has been completely split by the fusion of the enfolding cell membrane.

product rule a method used in CALCULUS to differentiate the product of two functions. If there are two functions, u and v, then the product function f(x) = uv can be differentiated using the following formula:

$$f'(x) = u'v + uv'$$

progesterone a STEROID hormone, which in mammals is important in pregnancy.

prophase the first stage of MEIOSIS or MITOSIS in cells of EUCARYOTES. During prophase, the CHROMOSOMES condense and can thus be studied using a microscope.

protein a complex, nitrogen-containing, organic compound of vital significance to all living matter. Proteins have high MOLECULAR WEIGHTS and comprise hundreds or thousands of AMINO ACIDS joined to form POLYPEPTIDE chains. The amino acid sequence confers upon each protein its particular properties. ENZYMES are another important group of proteins.

proton a particle that carries a positive charge and is found in the NUCLEUS of every ATOM. As an atom is electrically neutral, the number of protons equals the number of negatively charged ELECTRONS. Although the MASS of a proton (1.673×10^{-27} kg) is far greater than the mass of an electron (9.11×10^{-31} kg), their charges are equal in magnitude. The number of protons in the nucleus of an atom (ATOMIC NUMBER) is identical for any one element and is used to classify elements in the PERIODIC TABLE. For example, as every oxygen atom contains 6 protons, it has an atomic number of 6 in the periodic table, whereas every gold atom contains 79 protons and thus has an atomic number and periodic table position of 79.

pseudopodium (*plural* **pseudopodia**) the temporary projection from the body of certain cells. Pseudopodia are formed in simple, single-celled organisms, such as amoeba, as a mechanism for locomotion and food intake. They are also formed by white blood cells, which use PHAGOCYTOSIS to ingest particles.

PTFE (polytetrafluoroethene) a thermosetting PLASTIC produced by the polymerization of tetrafluoroethene (CF_2CF_2). Under its trade names of Teflon and Fluon, it is used to line saucepans, where its chemical unreactivity and heat resistance are useful. It is also used in engineering applications.

pulmonary artery one of the two arteries that carry deoxygenated blood from the HEART to the lungs, where it is oxygenated. The pulmonary arteries are the only ones that carry blood with a high concentration of carbon dioxide rather than a high concentration of oxygen. All other arteries carry oxygenated blood to the tissues, where oxygen is exchanged for carbon dioxide.

pulmonary vein one of the four veins that carry oxygenated blood from the lungs (two veins leave both the left and right lungs) to the left ATRIUM of the HEART. The pulmonary veins are unique, as they carry oxy-

pulsar

genated blood while all other veins carry deoxygenated blood back to the heart after it has exchanged oxygen for carbon dioxide in the various tissues of the body.

pulsar a star that is a sorce of radio frequency radiation (*see* ELECTROMAGNETIC WAVES) which is emitted in regular short bursts. Many have been located with radio telescopes, and it is thought that they are collapsed, rotating NEUTRON STARS.

purine one of the two different structures that form the base components of DNA and RNA. A purine has a double ring structure that consists of both carbon and nitrogen atoms. The bases, ADENINE and GUANINE, are both purines that will form hydrogen bonds with their complementary PYRIMIDINE bases to form the double helix of the DNA molecule.

PVA (polyvinyl acetate) a PLASTIC produced by the polymerization of vinyl acetate. It is used in coatings, adhesives and inks.

PVC (polyvinyl chloride or polychloroethene) the most widely used of the vinyl PLASTICS formed by POLYMERIZATION of vinyl chloride (chloroethene H_2CCHCl). PVC is used for pipes, ducts, mouldings and as a fabric in clothing and furnishings.

pyrimidine one of the two different structures that form the base components of DNA and RNA. A pyrimidine has a single ring, consisting of both carbon and nitrogen atoms. The bases, cytosine, THYMINE and URACIL, are all pyrimidines.

pyruvate a colourless liquid formed as a key intermediate in the metabolic process of GLYCOLYSIS and the production of ATP.

Pythagoras' theorem the geometrical theorem that states that in any right-angled triangle, the square of the HYPOTENUSE is equal to the sum of the squares of the two shorter sides. This theorem is named after the Greek philosopher and mathematician of the 4th century BC. For a given right-angled triangle in which the sides are x and y units long, the hypotenuse (h) can be obtained from $h^2 = x^2 + y^2$. Pythagoras' theorem provides a method of calculating the length of any side of a right-angled triangle if the lengths of the other two sides are known.

Q

quadratic equation an equation that has the general form $ax^2 + bx + c = 0$. The roots of any quadratic equation can be obtained from the formula:

$$x = \frac{-b \pm \sqrt{b^2 - 4ac}}{2a}$$

Part of this equation, $b^2 - 4ac$, is called the discriminant and describes the roots of a quadratic equation. If the discriminant has a positive value ($b^2 - 4ac > 0$) then the roots are real and distinct, but if the discriminant has a negative value ($b^2 - 4ac < 0$), then the roots are imaginary (graph does not cut x-axis). If the discriminant is zero ($b^2 - 4ac = 0$), then the roots are real but not distinct as both roots have the same value.

quadrilateral any geometric shape that has four sides. Some examples of quadrilaterals are the rhombus, kite, parallelogram and the rectangle.

quantum (*plural* **quanta**) a small, discrete quantity of radiant energy. Electromagnetic radiation (*see* ELECTROMAGNETIC WAVES) is explained in terms of small particles as well as waves, as it is assumed that it can be absorbed or emitted in quanta. The energy of one quantum (E) is derived from the equation, $E = hv$, where v is the FREQUENCY of the radiation and it is PLANCK'S CONSTANT.

quantum numbers a set of four numbers used to describe atomic structures (*see* ORBITALS). The first, n, the principal quantum number, defines the shells (stationary orbits in BOHR'S THEORY) which are visualized as orbitals. The orbit nearest the NUCLEUS has n = 1, and contains 2 ELECTRONS. The second shell, n = 2, contains 8 electrons, the maximum number of electrons in each shell being limited by the formula $2n^2$; the orbital quantum number, l, defines the shape of the orbits within one shell, which are designated s, p, d, f orbits; the magnetic orbital quantum number, m, which sets the spatial position of the orbit within a strong magnetic field; and s the spin quantum number, based upon the assumption that no two electrons may be exactly alike, and thus opposite spins are invoked for pairs of electrons.

quark any of the theoretical building blocks that participate in the strong interactive forces between ELEMENTARY PARTICLES. Originally, it was postulated that there were three types of quark, each carrying a charge that is less than that of one electron, but the need to explain new phenomena may make it necessary to have more types of quarks.

quartz one of the most widely distributed ROCK-forming minerals, SiO_2. It occurs in all kinds of rocks, and in its various crystalline forms and with certain impurities, it forms semi-precious stones, e.g. amethyst and agate. Quartz crystals exhibit the PIEZOELECTRIC EFFECT.

quasar "quasi-stellar" objects, which are extremely compact, light-emitting and yet enormously distant bodies—up to 10^{10} LIGHT YEARS away.

quinine a colourless ALKALOID with a very bitter taste, which was used in the treatment of malaria.

quotient the result obtained when a mathematical quantity (number, function or equation) is divided by another quantity.

quotient rule a mathematical method used in CALCULUS to differentiate the QUOTIENT of two functions. The quotient rule for the functions u and v, f(x) = u/v, is as follows:

$$f'(x) = \frac{u'v + uv'}{v^2}$$

R

radar (*acronym for* Radio Detection And Ranging) the use of radio waves to detect the presence and distance of objects. Used in navigation of aircraft, ships, missiles and SATELLITES.

radian an alternative to the degree in measuring angles, derived from the angle subtended at the centre of a circle by an arc equal to the length of the radius. There are $2\pi(6.284)$ radians in a full circle ($360°$).

radiation the emission of energy from a source, applied to ELECTROMAGNETIC WAVES (radio, light, X-rays, infrared, etc), particles (α, β, protons, etc), and sound.

radical a group of atoms (within a compound), usually unable to exist independently, which is unchanged in reactions affecting the rest of the molecule. (Now often referred to as a group.)

radical sign the inverse operation to forming a POWER is that of extracting a root and is expressed by a radical sign $\sqrt{}$, e.g. $\sqrt{4} = 2$

radioactivity the emission of α or β particles and/or γ rays by unstable elements, while undergoing spontaneous disintegration.

radiocarbon dating a method of dating organic material, although it is only applicable to the last 6000-

8000 years. There is a small proportion of radioactive ^{14}C in the atmosphere, which is taken up naturally by plants and animals. When an organism dies, the uptake ceases and the ^{14}C decays with a HALF-LIFE of 5730 years. Comparison of residual radioactivity with modern standards enables an age to be calculated for a sample.

radio (waves) ELECTROMAGNETIC radiation used to communicate through space without an intermediate physical link. The information thus conveyed can include sound, pictures, and digital data.

radiolysis the chemical decomposition of a substance subjected to ionizing radiation.

rainbow the characteristic display of colours formed by the REFRACTION and internal REFLECTION of sunlight by raindrops in the air.

Raoult's law a law that states that the vapour pressure of an (ideal) solution is the sum of the vapour pressures of each component.

rare earth elements *see* **lanthanides**.

rate of change (*also called* the DERIVATIVE) is the slope of a graph y = f(x) at a given point c, or in more precise terms, is the limit, as h approaches zero, of f(c + h) - f(c)/h.

ratio numbers these are used to compare the sizes of two or more quantities. If in a class of 24 pupils there are 8 boys and 16 girls, the ratio of boys to girls is:

8 : 16 or 1 : 2

It is usual to try to reduce one of the numbers to 1, and

rational number a number that can be obtained by dividing one quantity by another quantity: a/b with b = 0. This includes all whole numbers and most FRACTIONS.

rayon the term applied formerly to "artificial silk," but now to two manmade cellulose fibres, viscose and cellulose acetate rayon.

reagent a chemical substance or solution that is used to produce a characteristic reaction in chemical analysis.

real number any RATIONAL or IRRATIONAL number. Real numbers exclude imaginary numbers and COMPLEX NUMBERS.

recessive allele a gene form that is not expressed and will therefore not affect the PHENOTYPE of the organism unless the organism is HOMOZYGOUS for the recessive allele. Although an organism that is HETEROZYGOUS for a recessive allele will possess this allele, the dominant form of the gene will be expressed, thus masking the presence of the recessive form.

reciprocal the inverse ("other way up") of a FRACTION; the reciprocal of a number A is 1/A. For example, the reciprocal of 5/12 is 12/5, and the reciprocal of 6 is 1/6.

recombinant DNA a new DNA sequence formed by the insertion of a foreign DNA fragment into another DNA molecule. Recombinant DNA is used extensively throughout GENETIC ENGINEERING, when bacteria are frequently used as hosts for the expression of recombinant DNA molecules and the subsequent coding for the desired protein. It is particularly useful for producing a significant quantity of a human PROTEIN, such as INSULIN.

recombination *see* **genetic recombination**.

rectangle *see* **parallelogram**.

rectifier a device that converts ALTERNATING CURRENT into direct current.

red blood cell *see* **erythrocyte**.

redox potential a method for evaluating the REDUCTION or OXIDATION potential of a reactant. Redox potentials are arranged on an arbitrary scale, which uses the standard hydrogen electrode as the reference redox reaction by assigning it a potential of zero volts. The strongest REDUCING AGENTS, i.e. those most easily oxidized, are at the top of the scale while the strongest OXIDIZING AGENTS, i.e. those most easily reduced, are at the bottom of the scale.

redox reaction a chemical reaction in which both REDUCTION and OXIDATION are involved. If the overall REDOX POTENTIAL for such a reaction has a positive value, then it is a spontaneous and feasible reaction.

reducing agent any substance that will lose ELECTRONS during a chemical reaction. Reducing agents will readily cause the REDUCTION of other atoms, molecules or compounds, depending on the strength of the reducing agent and the reactivity of the other reactant. The strongest reducing agents are active alkali metals such as lithium (Li), potassium (K), barium (Ba), and calcium (Ca).

reduction any chemical reaction that is characterized by the loss of oxygen or the gain of ELECTRONS from one of the reactants. A molecule is also said to be reduced if it has gained a hydrogen atom. There is always simultaneous OXIDATION if reduction has occurred in any reaction.

reduction formulae equations that allow the TRIGONOMETRIC FUNCTION of any angle to be expressed as the trigonometric function of an ACUTE angle. Tables giving the values of trigonometric functions for angles at various intervals have been computed, so the problem of finding the trigonometric function of any angle is reduced to finding the trigonometric function of an angle between 0° and 90°, looking this up in the table, and then prefixing the proper sign.

reflection the property of certain surfaces whereby rays of light falling upon them are returned (reflected) in accordance with definite laws. The incoming or incident ray becomes the reflected ray.

refraction the bending of, most commonly, a ray of light, on travelling from one medium to another. The refraction occurs at the interface between the media and is caused by the light travelling at different velocities in different media. The incident ray becomes the refracted ray upon refraction (*see* REFRACTIVE INDEX).

refractive index (n) the ratio of the SINE of the angle of incidence (the angle betweeen the incident ray and

regeneration

the line drawn PERPENDICULAR to the surface at that point) to the sine of the angle of refraction, when light is refracted from a vacuum into the medium.

regeneration the repair or regrowth of bodily parts of an organism that have been damaged and subsequently lost. Regeneration is rare in higher, complex animals but is quite common in lower, simpler animals in which the extent of regeneration can range from limb regeneration in crustaceans to the regeneration of the whole organism from one segment, as in certain annelid worms. Regeneration is common in plants and occurs naturally, as in VEGETATIVE PROPAGATION, or can be induced to propagate plants of economic importance, such as the potato and tobacco plants. Complete regeneration of any plant is only possible if its vegetative cells have retained the full genetic potential (i.e. are TOTIPOTENT), enabling them to replicate every part of the plant.

regression in mathematics, the connection between the expected value of a random VARIABLE and the values of one or more possibly related variables. In biology, the tendency to an average state from an extreme one.

relative atomic mass (*formerly called* **atomic weight**) the mass of atoms of an element given in atomic mass units (u), where $1u = 1.660 \times 10^{-27}$ kg.

relativity the theory derived by EINSTEIN that establishes the concept of a four-dimensional space-time continuum where there is no clear demarcation between three-dimensional space and independent time, hence space and time are considered to be bound together. The important results of the theory include the appreciation that the mass of a body is a function of its speed; the derivation of the mass-energy equation, $E = mc^2$ (where c = speed of light), and the relative nature of time itself, i.e. there is no absolute value or interval of time.

resins natural resins are organic compounds secreted by plants and animals e.g. rosin, derived from pine trees. Synthetic resin is the term now applied to any synthetic PLASTIC material produced by polymerization.

resistance (R) measured in OHMs and calculated as the potential difference between the ends of a CONDUCTOR, divided by the CURRENT flowing. Superconductors apart, materials resist the flow of current to varying degrees, and some of the electrical energy is thereby converted to heat.

resistor a component of electric CIRCUITs, used to provide a known RESISTANCE.

resolution in chemistry, the separation of an optically inactive compound or mixture into its optically active components (*see* OPTICAL ISOMERISM).

resonance the creation of vibrations in a system by the application of a periodic force, e.g. from another vibrating system. As the FREQUENCY of the applied force becomes nearer to the natural frequency of the system, the vibrations increase, to reach a maximum when the two frequencies are equal.

respiration the process by which living cells of an organism release energy by breaking complex organic compounds into simpler ones using enzymes. Respiration can occur in the presence of oxygen (AEROBIC RESPIRATION) or in its absence (ANAEROBIC RESPIRATION) and has an initial stage called GLYCOLYSIS, which is common to both forms of respiration. The term respiration is also used, although less frequently, for gaseous exchange (better known as breathing) in an organism, which involves the uptake of oxygen from, and the release of carbon dioxide to, its surrounding environment.

retrovirus *see* **virus**.

rheostat a RESISTOR of variable RESISTANCE.

rhesus *see* **blood grouping**.

rhombus *see* **parallelogram**.

ribosomal RNA (rRNA) one of the three major classes of RNA, which is transcribed from DNA in a structure of eucaryotic nuclei called the NUCLEOLUS. Along with many PROTEINS, ribosomal RNA forms the cellular structures called RIBOSOMES, which are found in both EUCARYOTIC and PROCARYOTIC cells.

ribosome the cellular structure that is the site of PROTEIN synthesis in all EUCARYOTIC and PROCARYOTIC cells. Ribosomes are composed of one large and one small sub-unit, which contain RIBOSOMAL RNA and associated proteins. Analysis of procaryotic and eucaryotic ribosomes indicates that they share the same evolutionary origins as their structure, and the

RNA they contain (except a segment unique to eucaryotes) are virtually identical. Ribosomes assemble at one end of a MESSENGER RNA molecule and move along the molecule to build the POLYPEPTIDE chains of all proteins in a process called TRANSLATION.

right angle an angle of 90°.

ring compound *see* **closed-chain**.

RNA (*abbreviation for* **ribonucleic acid**) a NUCLEIC ACID that exists in all living cells. It is made up of a single-stranded chain of alternating ribose and phosphate units, BASE PAIRed between ADENINE and THYMINE or CYTOSINE and URACIL.

rock an aggregate of minerals or organic matter. Rocks are classified into three types: IGNEOUS, SEDIMENTARY, and METAMORPHIC.

röntgen (R) a radiological term defining the X-ray or gamma ray dose producing ions carrying a specific charge.

röntgen equivalent man a unit of radiation dose which has now been replaced by the SIEVERT.

rubber a natural hydrocarbon POLYMER (polyisoprene) from the *Hevea brasiliensis* tree. Items made from rubber are produced by adding various agents followed by vulcanization (heating in the presence of sulphur). Synthetic rubbers are polymers (or copolymers) of simple molecules.

ruminant any mammal that has four compartments in the stomach to aid the digestion of large amounts of plant matter. Ruminants (order Artiodactyla) include

ruminant

cattle, sheep, deer and giraffes. In the first section of their stomach, the rumen, food is enveloped in a mucus and is partially digested by an ENZYME called cellulase, which is supplied by the billions of bacteria living in the rumen. After this, the food is regurgitated to the mouth, and, after chewing, it passes through a further two sections (water is removed) and eventually ends up in the true stomach, the abomasum, which contains the enzymes essential for complete digestion.

S

saccharides SUGARS (and therefore CARBOHYDRATES) divided into mono-, di-, tri- and polysaccharides. Monosaccharides are the basic units, simple sugars; disaccharides, e.g. sucrose and lactose, are formed by condensing two monosaccharides and removing water. Sucrose gives, on HYDROLYSIS, a mixture of glucose and fructose; trisaccharides comprise three basic units and polysaccharides are a large class of natural carbohydrates including STARCH and CELLULOSE.

salient an interior angle in a polygon that is less than 180°.

salt a compound formed when a metal ATOM replaces one or more hydrogen atoms of an ACID (*see also* BASE).

saponification a process in which ESTERS are hydrolised (*see* HYDROLYSIS) by the action of acids, alkalis, boiling with water or superheated steam. If acids are used, it is the opposite process to esterification, but if alkalis are used then SOAPS result (hence the name).

saprotroph an organism that obtains its nutrition from dead and decaying organic matter. The group includes many bacteria and fungi, which are responsible for

satellite

the release of nitrogen, carbon dioxide and other nutrients from the decomposing matter.

satellite any body, whether natural or manmade, that orbits a much larger body under the force of gravitation. Hence the moon is a natural satellite of the earth.

saturated compound a group of compounds with no double or triple BONDS; i.e. they do not form ADDITION compounds through the joining of hydrogen atoms or their equivalent.

saturated solution a SOLUTION of a substance that exists in EQUILIBRIUM with excess SOLUTE present.

scalar a quantity that has MAGNITUDE but not direction; a REAL NUMBER, as opposed to a VECTOR. The magnitude of a vector two units in length is real number or scalar 2.

scalar product the SCALAR product of two VECTORS (written A.B) is the product of the REAL NUMBERS associated with them (their MAGNITUDES) and the COSINE of the angle between their two directions, i.e. $A.B = |A| |B| \cos \theta$.

scattering the dispersal of waves or particles upon impact with matter; applicable to light, atomic particles, etc.

scientific notation a useful method of writing large and small numbers. The scientific notation for a number is that number written as a power of 10 times another number, x, such that x is between 1 and 10 ($1 \leq x < 10$), e.g. $145,800 = 1.458 \times 10^5$.

scintillation small light flashes caused by ionizing

radiations (α, β, or γ rays) striking certain phosphors, i.e. substances that luminesce.

secant the function of an angle in a right-angled triangle, given by the reciprocal of the COSINE function: the secant of an angle A is 1/cosA.

second a unit of plane angle, equal to 1/60th of a minute or 1/3,600th of a degree, or π/648,000 radian. Also 1/86400th of the mean solar day.

sedimentary rock one of the three main ROCK types. Sedimentary rocks are formed from existing rocks through processes of erosion, denudation and subsequent deposition, compaction and cementation. The main types are: terrigenous—derived from existing rocks on the land, e.g. sandstones and shales; organic—produced by organic processes, e.g. limestones formed from coral reefs; chemical—precipitated from solution, e.g. evaporites such as gypsum; and volcanogenic—associated with volcanic action, e.g. volcanic ash deposited as tuff, or BENTONITE.

semiconductor an element or compound with average resistivity (RESISTANCE related to dimensions) between that of a CONDUCTOR and an INSULATOR. The commonest are silicon, germanium, selenium, and lead sulphide. These and other materials form the basis of TRANSISTORS, DIODES, thyristors, and integrated circuits. The transistor is the basis of electronic circuits and consists of minute slices of silicon in a sandwich, altered chemically to confer differing conductivities. The passage of current through the sand-

senescence

wich can be controlled by a weak current through the central slice, thus creating an electrically operated switch.

senescence the process of ageing that is characterized by the progressive deterioration of tissues and the metabolic functions of their cells. According to research, senescence may be caused by the accumulation of genetic mutations within the body's cells or the expression of undesirable GENES in the later years of an individual's life. Some organisms are able to suppress senescence by REGENERATION, a process common in many simple invertebrates and one achieved by some plants using VEGETATIVE PROPAGATION.

sequence a succession of mathematical entities, x_1, x_2, ..., x_n, ..., which is indexed by the positive integers. The term x_n is the nth term or general term. A sequence may be defined by stating its nth term, e.g. a sequence whose nth term is n/n + 1 is the sequence 1/2, 2/3, 3/4, 4/5, 5/6, ..., n/n + 1, ...

series an expansion of the form $x_1 + x_2 + x_3 + ..., + x_n + ...$, where xn are real or complex numbers.

series circuit where a common current flows through the components in a circuit.

serum *see* **blood serum**.

sex chromosome one of a pair of chromosomes that play a major role in determining the sex of the bearer, with a different combination in either sex. An individual is said to be homogametic when it has a HOMOL-

OGOUS pair of sex CHROMOSOMES (as in the XX of female mammals) and is said to be heterogametic when it has different sex chromosomes forming its pair (as in the XY of male mammals). Sex chromosomes contain GENES that decide an individual's sex by controlling the sexual characteristics of the individual, e.g. testes in human males and ovaries, breasts, etc in human females.

sex linkage the location of a GENE on a SEX CHROMOSOME, although the expression of the gene does not necessarily affect the sexual characteristics of the individual. Some examples of sex-linked genes include red-green colour blindness and haemophilia, both RECESSIVE genes found on the X-chromosomes. Such X-linked genes cannot be passed from father to son, as the father contributes only the Y-chromosome while the mother contributes the X-chromosome to a son. The Y-chromosome contains fewer specific genes than the X-chromosome other than those responsible for maleness, and any Y-linked genes will only be inherited by male offspring.

sexual reproduction the production of progeny that have initially arisen from the fusion of male and female GAMETES in a process called FERTILIZATION. In DIPLOID organisms, sexual reproduction must be preceded by MEIOSIS to form the HAPLOID gametes if there is not to be a doubling of the number of chromosomes in all sexually reproduced offspring. As sexual reproduction involves MEIOSIS, it introduces greater genetic variation in a species, because GENETIC RECOMBINATION can occur during meiosis, with the result that any offspring will have gene combinations that differ from its parents.

sievert the SI unit of radiation dose defined as that radiation delivered in one hour at a distance of one centimetre from a point source of one milligram of radium enclosed in platinum which is 0.5mm thick.

sigma the symbol Σ (Greek capital Sigma—"S" for "sum") used in statistics and mathematics. For example, Σx = the sum of all the values that x can assume, and Σx^2 = square each value of x then add the results.

sigma bond (σ) the bond type formed in a carbon-carbon single bond where two atomic ORBITALS overlap to form a molecular orbital surrounding the two carbon nuclei.

silicon chip a SEMICONDUCTOR chip of crystalline silicon onto which is printed a microelectronic (integrated) circuit for use in computers, radios, etc.

Simpson's rule (*also called* **parabolic rule**) a formula for approximating definite integrals, which states that the integral of a real-valued function $y = f(x)$ on an interval (a, b) is approximated by dividing the interval into an equal number of n parts at the points $x_1, x_2, ..., x_{n-1}$. The ordinates at these points are $y_1, y_2, ..., y_{n-1}$, and the width of the divisions $h = (b - a)/2$, so the area under the curve between a and b is given by $1/3\ h(y_a + 4y_1 + 2y_2 + 4y_3 + ... + 2y_{n-2} + 4y_{n-1} + y_b)$.

simultaneous equations two or more equations with

two or more unknown variables, which may have a unique solution, e.g. 4a - b = 10 and 3a + 2b = 24 yields the solution a = 4 and b = 6.

sine a function of an angle in a right-angled triangle, defined as the ratio of the length of the side opposite the angle to the length of the HYPOTENUSE.

sine rule in any triangle A/sin a = B/sin b = C/sin c; or, any side divided by the sine of the opposite angle is equal to any other side divided by the sine of its opposite angle.

single bond *see* σ **bond**.

SI units a system of coherent metric units—Système Internationale d'Unités (*see* APPENDIX 4).

skew the degree of asymmetry in a distribution curve. A skewed distribution has its modal value (the "hump" of the curve) either to the left or the right of the mean value. The degree of skewness can be measured by Pearson's COEFFICIENT of skewness, given by:

$$S = 3(\text{mean value - median value})/\sigma$$

where σ = standard deviation.

skew lines lines in space that are not parallel and do not intersect. Skew lines cannot be coplanar.

soap the sodium or potassium salts of the FATTY ACIDS, stearic, palmitic and oleic acid. Soaps are produced by the action of sodium or potassium hydroxide on fats (*see* SAPONIFICATION).

solenoid a tightly wound, cylindrical coil of wire, which generates a magnetic field when current is passed through the coil.

solid in geometry, a figure with the dimensions of length, width and breadth and thus a measurable VOLUME. In chemistry, a state of matter in which the component MOLECULES, ATOMS, or IONS sustain a constant position in relation to one another, i.e. they exhibit no translational motion.

solid state physics the study of all the properties of solid materials, but especially of SEMICONDUCTORS and "solid-state" devices, i.e. devices with no moving parts, as in integrated circuits, TRANSISTORS, etc.

solstice the time at which the sun reaches its most extreme position north or south of the equator. There are two such instants in the year.

solubility the concentration of a SATURATED SOLUTION is called the solubility of the given solute in the particular solvent used, measured in kgm^{-3}.

solute one substance dissolved in another. A solute dissolves in a SOLVENT to form a SOLUTION.

solution a single phase mixture of two or more components, which usually applies to solids in liquids and often refers to a solution in water (aqueous solution). However, other solutions include gases in liquids and liquids in liquids.

solvent a substance, usually a liquid, that can dissolve or form a SOLUTION with another substance.

somatic cell any of the cells of a multicellular organism (plant or animal) other than the reproductive cells (GAMETES).

sonar (*acronym for* Sound Navigation Ranging) a

device that transmits high frequency sound and collects returning sound waves that have been reflected from submerged objects. The depth is indicated by the time taken for the return journey.

sonic boom the loud bang created by shock waves from the leading and trailing edges of an aircraft travelling supersonically. The boom results from the aircraft overtaking the pressure waves it creates ahead of itself.

sound the effect upon the ear created by air vibrations with a frequency between 20Hz and 20kHz. More generally, mechanical vibrations and waves in gases, liquids and some solids.

Southern Lights *see* **aurora**.

specific heat capacity the heat required by unit mass to raise its temperature by one degree (SI units—joules per kg per Kelvin).

specific latent heat *see* **latent heat**.

spectral types a classification system for stars, based upon the SPECTRUM of light they emit. The sequence is, in order of descending temperature: O—hottest blue stars; B—hot blue stars; A—blue white stars; F—white stars; G—yellow stars; K—orange stars; M—coolest red stars.

spectrum the range of WAVELENGTHs obtained upon resolution of ELECTROMAGNETIC radiations. An obvious example is the coloured "rainbow" bands obtained when white light passes through a prism.

speed for a body moving in a straight line or contin-

speed of light

uous curve, the ratio of distance covered to the time required to cover that distance. Units vary, e.g. metres per second (ms^{-1}), miles per hour (mph).

speed of light as revealed in the theory of RELATIVITY, a universal and absolute (independent of the speed of the observer) value that is 2.998×10^8 ms^{-1}, or 186281 miles per second.

speed of sound the value for the speed (VELOCITY) of sound depends upon the nature of the medium and the temperature. In air at 0°C, the speed is 332 ms^{-1}, or about 760 mph. In fresh water, the speed is 1410 ms^{-1}.

sphere a circular solid figure with all points on its surface an equal distance from the centre. In two-dimensional CARTESIAN CO-ORDINATES, the equation of a sphere is $(x - a)^2 + (y - b)^2 + (z - c)^2 = r^2$. For a sphere of radius r, vol = $4/3\pi r^3$; surface area $A = 4\pi r^2$.

spherical co-ordinates (*also called* **spherical polar co-ordinates**) a three-dimensional polar co-ordinate system in which the position of a point P is given by the co-ordinates (r, θ, φ), where r is the radius vector with respect to the origin O of two axes at right angles; θ is the angle between the vertical (polar) axis and the radius vector; and φ is the angle between the horizontal axis (x-axis) and the projection of the radius vector on the horizontal plane.

spontaneous combustion the ignition of a substance without application of a flame. It may occur through the production of heat from slow OXIDATION within the substance.

standing wave

spontaneous generation a now discredited theory that living organisms could arise from non-life. It was believed that spontaneous generation could occur in, for example, rotting meat or fermenting broth, giving rise to an individual organism, but it is now known that all new organisms originate from the parent organism from whom they have inherited a genetic ancestry.

spore a small reproductive unit, usually consisting of one cell, that detaches from the parent and disperses to give rise to a new individual under favourable environmental conditions. Spores are particularly common in fungi and bacteria but also occur in all groups of green land plants such as ferns, horsetails and mosses.

square *see* **polygon**.

stalactite a hanging deposit of calcium carbonate ($CaCO_3$) formed from the roof of a cave by drips of calcium-rich solutions. Stalactites resemble icicles.

stalagmite an upstanding growth of calcium carbonate ($CaCO_3$) formed on the floor of a cave by drips of calcium-rich solutions. Often found with stalactites, when the two forms may eventually meet and join.

standing wave a disturbance produced when two similar wave motions are transmitted in opposite directions at the same time. This results in INTERFERENCE, with the combined wave effects producing maxima and minima over the area of interference. The resultant waveform is contained within fixed points and

starch

does not move, hence standing or stationary wave.

starch a polysaccharide (*see* SACCHARIDES) found in all green plants. It is built up of chains of GLUCOSE units arranged in two ways, as amylose (long unbranched chains) and amylopectin (long cross-linked chains). Potato and some cereal starches contain about 20-30 per cent amylose and 70-80 per cent amylopectin.

static electricity stationary electric charges which result from the electrostatic field produced by the charge (*see also* VAN DER GRAAF GENERATOR).

statistics the branch of mathematical science dealing with the collection, analysis and presentation of quantitive data, and drawing inferences from data samples by the use of probability theory.

stereochemistry the part of chemistry that covers the spatial arrangement of atoms within a molecule (*see* ISOMER).

stereoisomerism isomerism due to the different arrangement in space of atoms within a molecule, giving ISOMERs that are mirror images of each other.

steroids a group of LIPIDS with a characteristic structure comprising four carbon rings fused together. The group includes the sterols (e.g. CHOLESTEROL), the BILE acids, some HORMONES, and vitamin D. Synthetic steroids act like steroid hormones and include derivatives of the glucocorticoids used as anti-inflammatory agents in the treatment of rheumatoid arthritis; oral contraceptives which are commonly mixtures of OESTRO-

GEN and a derivative of PROGESTERONE (both female sex hormones); anabolic steroids e.g. TESTOSTERONE, the male sex hormone, which is used to treat medical conditions such as osteoporosis and wasting. However, much publicity surrounds the use of the anabolic steroids by athletes, contrary to the rules of sports-governing bodies, to increase muscle bulk and body weight.

stoichiometry an aspect of chemistry that deals with the proportion of elements (or chemical equivalents) making pure compounds.

strain when forces acting upon a material produce distortion, it is said to be strained, or in a state of strain. Strain is represented as a ratio of the change in dimension or volume to the original dimension or volume.

stress force per unit area. When applied to a material, a corresponding STRAIN is created. The two main types are tensile (or compressive) and shear stress. Units, typically, include: kNm^{-2}, $lbfin^{-2}$.

stroboscope an instrument used to view rapidly moving objects by shining a flashing light source, of variable periodicity, at the object. Synchronization of the two frequencies renders the object apparently immobile.

stroma (*plural* **stromata**) any tissue that functions as a framework in plant cells (*see* CALVIN CYCLE).

structural formula a formula providing information on the ATOMS present in a MOLECULE and the way that they are bound together, i.e. an indication of the structure.

strychnine

strychnine a crystalline ALKALOID with a very strong bitter taste and a very dangerous action upon the nervous system.

sublimation the production of a vapour directly from a solid, without going through the liquid phase.

substitution (reaction) a reaction in which an atom or group in a molecule is replaced by another atom or group, often hydrogen by a HALOGEN, hydroxyl, etc.

substrate in biology (1) the surface upon which an organism lives and from which it may derive its food. (2) a substance/reactant in a reaction which is catalysed by an ENZYME. In electronics, the single crystal of SEMICONDUCTOR used as the base upon which an integrated circuit or TRANSISTOR is printed.

sucrose a disaccharide CARBOHYDRATE ($C_{12}H_{22}O_{11}$) occurring in beet, sugar cane and other plants (*see* SACCHARIDES).

sugar a crystalline monosaccharide or oligosaccharide (a small number, usually two to ten, monosaccharides linked together, with the loss of water), soluble in water. The common name for SUCROSE.

sulphuric acid (H_2SO_4) a strong acid that is highly corrosive and reacts violently with water, with the generation of heat. It is manufactured by the CONTACT PROCESS. It is used widely in industry in the manufacture of dyestuffs, explosives, other acids, fertilizers, and many other products.

summation convention an abbreviated notation, used particularly in tensor analysis and RELATIVITY THEORY, in which a product of tensors is to be summed over all possible values of any index that appears twice in the expression.

sun the star nearest to earth and around which the earth and the other planets rotate in elliptical orbits. The sun has a diameter of 1.392×10^6 kilometres and its mass is approximately 2×10^{30} kilograms. The interior reaches a temperature of 13 million degrees Centigrade, while the visible surface is about 6000°C. The internal temperature is such that thermonuclear reactions occur, converting HYDROGEN to HELIUM with the release of vast quantities of energy. The sun is approximately 90 per cent hydrogen, 8 per cent helium, and is 5 million years old—roughly halfway through its anticipated life cycle.

sunspots the appearance of dark areas on the surface of the SUN. The occurrence reaches a maximum about every eleven years. Sunspots have intense magnetic fields and are associated with magnetic storms and effects such as the *aurora borealis* (*see* AURORA). The black appearance is due to the sunspots dropping in temperature, to about 4000K.

supernova (*plural* **supernovae**) a star which explodes, it is thought, due to the exhaustion of its hydrogen (*see* SUN), whereupon it collapses, generating high temperatures and triggering thermonuclear reactions. A large part of its matter is flung into space, leaving a residue that is termed a WHITE DWARF star. Such events are very rare, but at the time of explosion, the stars

superconductivity

become one hundred million times brighter than the sun.

superconductivity the property of some metals and alloys whereby their electrical RESISTANCE becomes very small around ABSOLUTE ZERO. The potential uses for **superconductors** include circuits in large computers, where superconductive circuits would generate far less heat than is the case presently. This would enable larger and faster computers to be built. Transmission of electricity and reduction of the associated heat loss is also under study.

superposable denoting a property of a geometrical figure that allows its co-ordinates to be transposed to coincide with those of another figure.

supersaturation the state of a SOLUTION when it contains more dissolved SOLUTE than is required to produce a SATURATED SOLUTION.

surd the root of quantity that can never be exactly expressed because it is an IRRATIONAL NUMBER, e.g. 3 = 1.73205…

surface integral the integral of a function of several variables with respect to the surface area over a surface in the domain of a function.

surface tension a tension created by forces of attraction between molecules in a liquid, resulting in an apparent elastic membrane over the surface of the liquid.

surfactant (*also called* **surface-active agent**) a compound that reduces the SURFACE TENSION of its SOLVENT, e.g. a detergent in water.

symbiosis a relationship between organisms, usually two different species, which has beneficial consequences for at least one of the organisms. There are various forms of symbiosis, including commensalism, where one party benefits but the other remains unharmed, and parasitism, where one party greatly benefits (the PARASITE) at the other party's expense. Symbiosis can also solely refer to mutualism, where both parties benefit and neither is harmed.

symbol a letter or letters that represent an element or an atom of an element (*see* Element Table in APPENDIX 2 for list of symbols).

symmetry the property of a geometrical figure whose points have corresponding points reflected in a given line (axis of symmetry), point (centre of symmetry, e.g. a circle) or plane (reflection).

synapse the area where one nerve cell makes contact with another. Nerve impulses pass across the small gap, from one cell to the other, by means of chemicals called transmitter substances.

synthesis the formation of a compound from its constituent elements or simple compounds.

systems analysis the application of mathematics to the analysis of a particular system, involving the construction of a mathematical model, which is then analysed and the results applied to the original system.

systolic blood pressure the pressure generated by the left VENTRICLE of the HEART at the peak of its contraction. Since the left ventricle has to pump blood to

systolic blood pressure

all parts of the body, it generates a higher pressure than the right ventricle, which pumps blood only to the lungs. In normal people, the systolic blood pressure is 120 mm of mercury (120 mm Hg), and when the ventricle relaxes, pressure is still maintained in the blood vessels. This resting pressure is called the diastolic pressure and is approximately 80 mm Hg. This is the familiar blood pressure measurement and is represented as 120/80. This fluctuation in pressure is responsible for the pulse, which also represents the heartbeat.

T

tangent a function of an angle in a right-angled triangle, defined as the ratio of the side opposite the angle to the length of the side adjacent to it. In geometry, a straight line that just touches the circumference of a circle.

tangent rule tan 1/2 (A - B)/tan 1/2 (A + B) = a - b/a + b.

tannic acid a polymeric ESTER-like derivative of gallic acid and GLUCOSE. It is used as a MORDANT in dyeing and in tanning and ink manufacture.

tartaric acid an organic acid which exists in four forms that are STEREOISOMERS. The commonest form (d-tartaric acid) is used in dyeing, the manufacture of baking powder and "health salts." Another form (dl-tartaric acid, or racemic acid) occurs in grapes.

taxis (*plural* taxes) the movement of a cell or organism in response to a stimulus in environment. This stimulus may be temperature (thermotaxis), light (phototaxis), gravity (geotaxis) or chemical (CHEMOTAXIS).

taxonomy the study, identification and organization of organisms into a hierarchy of diversity according to their similarities and differences. Taxonomy is con-

T-cell

cerned with the classification of all organisms, whether plant or animal, dead or alive, e.g. fossils are also important. Modern taxonomy provides a convenient method of identification and classification of organisms, which expresses the evolutionary relationships to one another.

T-cell a type of white blood cell (LYMPHOCYTE) that differentiates in a gland called the thymus, situated in the thorax. There are a whole variety of T-cells involved in the recognition of a specific foreign body (ANTIGEN), and they are particularly important in combating viral infections and destroying bacteria that have penetrated the cells of the body.

telophase the last stage of MEIOSIS or MITOSIS in EUCARYOTIC cells. During telophase, a nuclear membrane forms round each of the two sets of CHROMOSOMES that have formed separate groups at the spindle poles. The chromosomes decondense, the nucleoli reappear, and the cell eventually splits to form two daughter cells.

temperature degree of heat or cold against a standard scale.

terminal velocity the constant VELOCITY achieved by a body falling through a medium when the pull of GRAVITY is equalled by the frictional resistance.

terpenes colourless, liquid hydrocarbons occurring in many fragrant natural oils of plants. The general formula is $(C_5H_8)n$ where C_5H_8 is the basic isoprene unit. This leads to their classification: monoterpenes are

$C_{10}H_{16}$; sesquiterpenes $C_{15}H_{24}$; diterpenes $C_{20}H_{32}$, and so on.

testosterone a male sex hormone that promotes the development of male characteristics.

thermistor a temperature-sensitive SEMICONDUCTOR whose RESISTANCE decreases with temperature increase. Thermistors are used for temperature measurement and compensation.

thermocouple a device for measuring temperature, comprising two metallic wires joined at each end. The temperature is measured at one join, and the other join is kept at a fixed temperature. A temperature difference between the two joins creates a thermoelectric e.m.f. (electromotive force = VOLTAGE), which causes a current to flow. Either the voltage or the current can be measured, thus creating a calibrated device.

thermodynamics the study of laws affecting processes that involve heat changes and energy transfer. There are essentially three laws of thermodynamics. The first (the law of conservation of energy) states that within a system, energy can neither be created nor destroyed. The second law says that the ENTROPY of a closed system increases with time. The result of the third law (or NERNST heat theorem) is that absolute zero can never be attained.

thermography the medical scanning technique whereby the INFRARED RADIATION or radiant heat emitted by the skin is photographed, using special film, to create images. An increase in heat emission signifies an increase in blood supply, which may be indicative of a CANCER. The technique is used to detect cancers, especially of the breast.

thermoluminescence a phenomenon whereby a material emits light upon heating due to ELECTRONS being freed from DEFECTS in crystals. The defects are generally due to ionizing radiations, and the principle is applied in dating archaeological remains, especially ceramics, on the assumption that the number of trapped electrons, caused by exposure to radiations, is a function of time. Although this is not absolutely correct, an estimate of the age of a piece of pottery can be obtained by heating and comparing the thermoluminescence with that of an item of known age.

thermometer an instrument used to measure temperature. Any property of a substance, providing it varies reliably with temperature, may form the basis of a thermometer. This includes expansion of liquids or gases, or changes in electrical RESISTANCE.

thermostat a device for maintaining a constant temperature through the supply or non-supply of heat when the required temperature is not achieved.

thixotropy the property of some fluids to be very viscous until a STRESS is applied, when the fluid flows more easily. This principle is utilized in non-drip paints.

thymine ($C_5H_6N_2O_2$) a nitrogenous base component of DNA that has the structure of a PYRIMIDINE. Thymine always base-pairs with ADENINE in a DNA molecule,

totipotency

but in RNA molecules it is replaced with URACIL.

thyroid gland *see* **endocrine system**.

tin a soft, malleable and ductile metal (SYMBOL Sn) which exhibits ALLOTROPY. It occurs naturally as oxides and is used to coat steel and in producing alloys (solders, fusible alloys, etc.).

titanium (Ti) a malleable and ductile metal that resembles iron. The main source is the ore rutile (TiO_2). It is characterized by lightness, strength and high resistance to corrosion. It is therefore useful in aircraft and missile manufacture.

titration the laboratory procedure of adding measured amounts of a SOLUTION to a known volume of a second solution until the chemical reaction between them is complete, enabling the unknown strength of one solution to be determined.

tomography a scanning technique that uses X-rays for photographing particular "slices" of the body. A special scanning machine rotates around the horizontal patient, taking measurements every few degrees over 180°. The scanner's own computer builds up a three-dimensional image that can then be used for diagnosis. Such a technique has the dual benefit of providing more detail than a conventional X-ray and yet delivers only one fifth of the dose.

tonne a metric ton (1000kg).

totipotency the capacity of a cell to generate all the characteristics of the adult organism. Totipotent cells have full genetic potential, unlike most adult cells,

transcription

which have lost this during the process of differentiation when they form cells with specialized functions.

transcription the formation of an RNA molecule from one strand of a DNA molecule. Transcription involves many processes, starting with the unwinding of the double-stranded DNA helix, along which an enzyme, called RNA polymerase, travels and catalyses the formation of the RNA molecule by pairing NUCLEOTIDES with the corresponding sequence of the DNA strand. As the RNA molecule leaves, the DNA reforms its double-stranded helix.

transducer a device that converts one form of energy into another, often a physical quantity into an electrical signal, as in microphones and photocells. The reverse also applies, as in loudspeakers.

transfer RNA (tRNA) one of the three major classes of RNA that functions as the carrier of AMINO ACIDS to RIBOSOMES, where the POLYPEPTIDE chains of PROTEINS are formed. Every tRNA molecule has a structure that will accept only the specific attachment of one amino acid.

transformation a rearrangement of the term of a mathematical expression. The changing (or mapping) of one shape into another by reflection, rotation, dilation, translation, etc.

transformer a device for changing the VOLTAGE of an ALTERNATING CURRENT. The unit consists essentially of an iron core with two coils of wire. Current fed into the primary coil generates a current in the secondary through ELECTROMAGNETIC INDUCTION. The ratio of the voltage between the coils is determined by the ratio of the number of turns in each coil.

transition point the point at which a substance may exist in more than one solid form, in equilibrium.

transition element an ELEMENT characterized by an incomplete inner electron shell and a variable valency. Metallic in nature, the chemical properties of one element resemble those of the adjacent element in the PERIODIC TABLE.

transitive a relation between mathematical entities such that if one object bears a relation to a second object, which bears a relation to a third object, then the first object bears this relationship to the third object, e.g. if a = b and b = c then a = c.

transistor a SEMICONDUCTOR device that is used in three main ways: as a switch; a rectifier (or DIODE, which conducts current in one direction, thus turning AC into DC); and as an amplifier creating strong signals from weak ones.

translation the synthesis of PROTEINS in a RIBOSOME that has MESSENGER RNA (mRNA) attached to it. As the mRNA molecule moves through part of the ribosome, a TRANSFER RNA molecule carrying the appropriate AMINO ACID will enter a site on the ribosome and will be released after it has contributed a new amino acid to the growing chain.

transpiration the loss of water vapour from pores (stomata) in the leaves of plants. Transpiration can sometimes account for the loss of over one sixth of the water that has been taken up by the plant roots. The transpiration rate is affected by many environmental factors—temperature, light and carbon dioxide (CO_2) levels, air currents, humidity and the water supply from the plant roots. The greatest transpiration rate will occur if a plant is photosynthesizing (*see* PHOTOSYNTHESIS) in warm, dry and windy conditions.

transpose in mathematics, to move one term or element from one side of an equation to the other with a corresponding reversal in sign. A MATRIX formed from another by interchanging the rows and columns: the transpose of matrix A is usually denoted AT.

transverse wave a wave in which the vibration occurs at right angles to the direction of wave propagation, e.g. an ELECTROMAGNETIC WAVE or, more simply, a wave on a taut piece of string.

trapezium a QUADRILATERAL with two parallel sides (*see also* ISOSCELES).

tribology the study of FRICTION, lubrication and wear, as when two surfaces are in contact in relative motion. It includes the study of substances that diminish wear, overheating, etc. in such circumstances.

trigonometric function one of the functions, such as $\sin(x)$, $\tan(x)$ and $\cos(x)$, obtained from studying certain ratios of the sides of a right-angled triangle.

trigonometry the study of right-angled triangles and their TRIGONOMETRIC FUNCTIONS.

trisomy the abnormal condition in which an organism

has three CHROMOSOMES rather than the normal pair for one type of chromosome. Trisomy can occur in humans and results in offspring with abnormal characteristics and shorter-than-average lifespans. One common example of trisomy is Down's syndrome, caused by the presence of three instead of two chromosomes of the number 21 type (all the other chromosomes are in normal pairs).

tungsten a hard grey metal used in alloys where its hardness and resistance to corrosion are valued. It is used in CARBIDE tools and electric lamp filaments. Tungsten carbide is almost as hard as diamond and is used extensively in abrasives.

U

ultracentrifuge a machine that generates high centrifugal forces as its rotor is capable of spinning at speeds of up to 50,000 revolutions per minute. The ultracentrifuge is most commonly used during the separation of the various ORGANELLES within cells. The larger and more dense organelles will form a deposit in the centrifuge tube more readily than the smaller, less dense ones. Thus the largest organelle of any normal cell, the NUCLEUS, will be deposited at the bottom of a centrifuge tube when the ultracentrifuge spins at a force of 600g for 10 minutes, whereas the smaller MITOCHONDRION needs a higher speed of 15,000g for 5 minutes to be deposited.

ultrasonic a term used to describe sound waves that are inaudible to humans as they have a frequency above 20kHz. Although the human ear is incapable of detecting such a high FREQUENCY, some animals, such as dogs and bats, can detect ultrasonic waves (*also known as* **ultrasound**). Ultrasound is used widely in industry, medicine and research. For example, it is used to detect faults or cracks in underground pipes and to destroy kidney stones and gallstones. The most recent development is its use in chemical processes to trigger reactions involved in the production of food, plastics and antibiotics.

ultraviolet radiation a form of radiation that occurs beyond the violet end of the visible light spectrum of ELECTROMAGNETIC WAVES. Ultraviolet rays have a FREQUENCY ranging from 10^{15}Hz to 10^{18}Hz, with a wavelength ranging from 10^{-7}m to 10^{-10}m. They are part of natural sunlight and are also emitted by white-hot objects (as opposed to red-hot objects, which emit INFRARED RADIATION). As well as affecting photographic film and causing certain minerals to fluoresce, ultraviolet radiation will rapidly destroy bacteria. Although ultraviolet rays in sunlight will convert steroids in human skin to vitamin D (essential for healthy bone growth), an excess can cause irreversible damage to the skin and eyes and damage the structure of the DNA in cells by producing THYMINE-thymine DIMERS. Fortunately, a great deal of the ultraviolet radiation from the sun is prevented from reaching the earth as the OZONE LAYER in the upper atmosphere acts as a UV filter.

unit cell the smallest fragment of a crystal that will reproduce the original crystal if the unit cells are arranged in a repeating, three dimensional pattern.

unit vector a VECTOR with a magnitude of one.

unity the number or numeral one; a quantity assuming the value of one.

universal gas equation *see* **gas laws**.

universal indicator a mixture of certain substances, which will change colour to reflect the changing pH

unsaturated

of a SOLUTION. Universal indicator is available in the form of a solution or paper strip and is used as an approximate measure of the pH of a solution by using the following chart as a guide:

Colour of indicator	red orange yellow green blue purple
pH	1 2 3 4 5 6 7 8 9 10 11 12 13 14

unsaturated a chemical term used to indicate that a compound or solution is capable of undergoing a chemical reaction due to specific physical properties of the compound or solution. In the case of unsaturated organic compounds, the carbon atoms are unsaturated as they form double or triple bonds and are thus capable of undergoing ADDITION REACTIONS. If a SOLUTION is described as unsaturated, then it contains a lower concentration of SOLUTE dissolved in a definite amount of the SOLVENT than the concentration of solute needed to establish the EQUILIBRIUM found in a SATURATED solution.

uracil ($C_4H_4N_2O_2$) a nitrogenous base component of the NUCLEIC ACID, RNA, that has the structure of a PYRIMIDINE. During TRANSCRIPTION, uracil will always form a BASE PAIR with ADENINE of the DNA template, and during TRANSLATION, uracil will always base pair with adenine of the MESSENGER RNA (mRNA) molecule.

uranium a metallic element that is radioactive and has the greatest mass of all naturally occurring elements (atomic mass of 238). Uranium has 92 protons within its nucleus and exists as three ISOTOPES, ^{238}U, ^{235}U, and ^{234}U—each of which undergoes ALPHA DECAY. Uranium will naturally disintegrate and pass through a series of other elements to form eventually a stable isotope of the element lead. When uranium is bombarded with NEUTRONS, however, it undergoes artificial disintegration to form two other heavy nuclei, releasing a very large amount of energy—this is the basic process underlying nuclear FISSION, which is used to generate energy in nuclear power stations. Uranium is also used in atomic bombs. Indeed, the first atomic bomb, dropped on Hiroshima in 1945, is believed to have contained two or more small quantities of the isotope ^{235}U, which were suddenly brought together by a device and the CHAIN REACTION of nuclear fission immediately ensued.

urea an organic molecule, $CO(NH_2)_2$, that is a metabolic byproduct of the chemical breakdown of PROTEIN in mammals. In humans, 20-30 grams of urea are excreted daily in the urine, and although urea is not poisonous in itself, an excess of it in the blood implies a defective kidney, which will cause an excess of other, possibly poisonous, waste products.

uridine a molecule consisting of the nitrogenous base URACIL and the ribose sugar that is a basic unit of RNA structure when a phosphate group (H_3PO_4) is added to form a NUCLEOTIDE.

V

vaccine a modified preparation of a VIRUS or BACTERIA that is no longer dangerous but is capable of stimulating an immune response and thus confers immunity against infection with the actual disease. Vaccines can be administered orally or by a hypodermic syringe and are not effective immediately as it takes time for the recipient's IMMUNE SYSTEM to develop a memory for the modified virus or bacterium by producing specific ANTIBODIES.

vacuum in theory, a space in which there is no matter. However, a perfect vacuum is unobtainable and the term describes a gas at a very low pressure.

vacuum tube *see* **diode**.

vagus nerve an important part of the nervous system that arises from the brain stem and runs down either side of the neck. The vagus nerves accompany the major blood vessels of the neck (internal jugular veins) and innervate the heart and other viscera in the chest and abdominal cavities. They are partially responsible for the control of the heart rate and other vital functions. Sudden stimulation of the vagus nerves can cause sudden death due to instant cardiac failure, with the victim suddenly dropping dead. Some examples of sudden death by vagal stimulation can include a blow in the solar plexus, any form of pressure on the neck, and sudden dilatation of the neck of the womb (e.g. illegal abortion).

valency the combining power or bonding potential of an ATOM or GROUP, measured by the number of hydrogen ions (H^+, valency 1) that the atom could replace or combine with. In an IONIC compound, the charge on each ion represents the valency e.g. in NaCl, both Na^+ and Cl^- have a valency of one. In COVALENT compounds, the valency is represented by the number of bonds that are formed, thus in carbon dioxide (CO_2), the carbon has a valency of 4 and oxygen 2 (*see* VALENCY ELECTRONS).

valency electrons the electrons present in the outermost shell of an atom of an element. Some elements always have the same number of valence electrons, e.g. hydrogen has one, ordinary oxygen has two, and calcium has two. The valence electrons of an atom are the ones involved in forming bonds with other atoms and are therefore shared, lost or gained when a compound or ION is formed.

valve a piece of tissue attached to the wall of a tube that restricts the flow of the blood being carried in one direction. The most important valves are the ones found in the HEART and VEINS.

Van der Graaf generator a machine that continuously separates electrostatic charges and in so doing produces a very high voltage. The fundamental structure of a Van der Graaf generator consists of a hollow metal sphere supported on an insulating tube. A motor-driven belt of, say, rubber or silk, carries positive charge from an electrode at the bottom of the belt into the sphere. The sphere gradually becomes positively charged, and in some generators of this type, voltages as high as 500kV or 10,000,000 volts can be produced. When used in conjunction with high voltage X-ray tubes, large machines with elaborate electrode systems can generate electrical energy, which is used to split atoms for research purposes.

Van der Waals' forces the weak, attractive force between two neighbouring atoms. They are named after the Dutch physicist, Johannes van der Waals (1837-1923), who first discovered the phenomenon. In any atom, the electrons are continually moving and therefore have random distribution within the electron cloud of the atom. At any one moment, the electron cloud of an atom may be distorted so that a transient DIPOLE is produced. If two non-covalently bonded atoms are close enough together, the transient dipole in one atom will disturb the electron cloud of the other. This disturbance will create a transient dipole in the second atom, which will in turn attract the dipole in the first. It is the interaction between these transient dipoles that results in weak, non-specific Van der Waals' forces. They occur between all types of molecules, but they decrease in strength with increasing distance between the atoms or molecules.

vaporization *see* **evaporation**.

variable a changing quantity that can have different values, as opposed to a constant. In the equation y = $3x^2 + 7$, x and y are variables, whilst 3 and 7 are constants. An INDEPENDENT VARIABLE is a variable in a function that determines the value of the other variable. A DEPENDENT VARIABLE has its value determined by other variables. So in this example, x is the independent variable and y is the dependent variable.

vector any physical quantity that has both direction and magnitude. Vectors include displacement, velocity, acceleration and momentum. In biology, the term vector represents the plasmid used to carry a DNA segment into the host's cells or the organism that acts as a mechanism for transmitting a parasitic disease, e.g. mosquitoes are vectors of malaria.

vegetative propagation a type of reproduction in which the non-sexual organs of the plant are capable of producing progeny. Vegetative propagation occurs naturally in certain plants, e.g. potato tubers and strawberry runners.

vein any thin-walled vessel that carries blood back from the body to the HEART. Veins contain few muscle fibres but have one-way VALVEs that prevent backflow, thus enabling the blood to flow from body areas back to the heart.

velocity the rate of change of position of any object. Velocity (v) is a VECTOR quantity, and therefore it should be expressed in both magnitude and direction.

velocity of light

The unit of velocity is metres per second (ms^{-1}) and can be calculated using the displacement (s) and time elapsed (t) as follows: $V = s/t$. An object is described as moving with constant velocity when it is travelling along a straight line in equal proportions of distance against time. However, it is more likely that an object's velocity changes with time, in which case the object is said to be accelerating (*see* ACCELERATION).

velocity of light the VECTOR quantity for light travelling through a given medium. The velocity of light in a vacuum and in air hardly differs and is approximately 3.0×10^8 ms^{-1}. However, the velocity of light in water is approximately 2.3×10^8 ms^{-1}, and it is this difference in velocity in air and water that explains the REFRACTION of light when passing from one medium to another (*see also* SPEED OF LIGHT).

vena cava one of the two major veins that empty into the right chamber of the HEART. The superior vena cava (SVC) carries the blood collected from the upper part of the body, e.g. neck and brain, while the inferior vena cava (IVC) carries the blood from the lower half of the body, e.g. liver, kidney and legs.

ventricle a major chamber of the HEART, which is thick-walled and muscular as it is the main pumping chamber. The outflow of the right ventricle is known as the PULMONARY ARTERY, which distributes blood to the lungs, and the outflow of the left ventricle is called the AORTA, which distributes blood to the head and the rest of the body.

vertex the point at which two sides of a polygon or the planes of a solid intersect.

virus the smallest microbe, which is completely parasitic as it is incapable of growing or reproducing outside the cells of its host. Most, but not all, viruses cause disease in plant, animal and even bacterial cells. Viruses are classified according to their nucleic acids and can contain double-stranded (DS) or single-stranded (SS) DNA or RNA. In infection, any virus must first bind to the host cells and then penetrate to release the viral DNA or RNA. The viral DNA or RNA then takes control of the cell's metabolic machinery to replicate itself, form new viruses, and then release the mature virus by either budding from the cell wall or rupturing and hence killing the cell. Some familiar examples of virus-induced diseases are herpes (double-stranded DNA), influenza (single-stranded RNA) and the retroviruses (single-stranded RNA, believed to cause AIDS and perhaps CANCER).

viscosity a property of fluids that indicates their resistance to flow. For example, oil is more viscous than water, and an object falling through oil is much slower than the same object falling through water because of the greater viscous force acting on it. A perfect fluid would be non-viscous.

vitamins organic compounds that are required in small amounts in the diet, to maintain good health. Deficiencies lead to specific diseases. Vitamins form two groups: A, D, E and K are fat-soluble, while C

viviparous

(ASORBIC ACID) and B (thiamine) are water-soluble.

viviparous a term describing any animal that gives birth to young that have developed inside its body. Viviparity also applies to some species of insect, e.g. the mite *Acarophenox*, whose young develop by devouring and thus killing the mother.

volatile a term describing any substance that can easily change from the solid or liquid state to its vapour.

volt (V) the unit of POTENTIAL DIFFERENCE. One volt is equal to one joule per coulomb of charge, i.e. $V = JC^{-1}$.

voltage the electrical energy that moves charge around a CIRCUIT. Voltage is the same as POTENTIAL DIFFERENCE, and is thus measured in VOLTs. It is calculated between two given points on a circuit, and can be derived from the following equation: $V = d \times E$, where V = potential difference, d = difference between 2 points, E = electric field strength.

volume the space occupied by any object or substance. The volume of a liquid will depend on the amount of container space it occupies, but the volume of any gas will vary with pressure and temperature. Volume is measured in cm^3 or m^3. The volume of a cube, cuboid or cylinder is equal to the area of the base × height; the volume of a pyramid or cone is equal to 1/3 of the area of the base × height. The volume of a sphere is $4/3\pi r^3$.

vulgar fraction an ordinary fraction with one number over the other, e.g. 3/5, 5/9, 1/16.

W

water potential the tendency of water to move by diffusion, osmosis or as vapour. At a pressure of one atmosphere, pure water is given a water potential value of zero, and hence all cells that water enters by osmosis have a water potential value less than zero. The water potential of any cell can be calculated using the following:

water potential = osmotic potential + pressure potential.

Watson, James Dewey (1928-) an American molecular biologist who, along with his colleague, the English biologist Francis H. Crick (1916-), constructed a model revealing the structure of the DNA molecule. In 1962, they shared the Nobel prize for their work, and their double-helical model of DNA, showing a simple, repeating pattern of paired nucleic acid bases, suggested a means by which DNA replicates. Watson published an account of the discovery of DNA structure in his book, *The Double Helix* (1968).

watt (W) a unit of power that is the rate of WORK done at 1 JOULE per second, i.e. $1W = 1Js^{-1}$.

wave a mechanism of energy transfer through a

wavelength

medium. The origin of the wave is vibrating particles, which store and release energy while their mean position remains constant as it is only the wave that travels. Waves can be classified as being either LONGITUDINAL WAVES, e.g. sound, or TRANSVERSE WAVES, e.g. light, depending on the direction of their vibrations. There is a basic wave equation that relates the wavelength (λ), frequency (f), and speed (c) of the wave as $c = f\lambda$. All forms of waves have the following properties: diffraction; interference; reflection and refraction. ELECTROMAGNETIC WAVES have all of these properties but differ from ordinary waves, such as water waves, in that they can travel through a vacuum, e.g. outer space. All travelling waves have the following equation:

$$y = y_0 \sin 2\pi \left(\frac{t}{T} - \frac{x}{\lambda}\right)$$

where y = displacemcent, y_0 = amplitude, t = time, T = 1/f (f=frequency), x = distance, λ = wavelength.

wavelength (λ) the distance between two similar points on a wave, which have exactly the same displacement value from the rest position. An example would be the distance between two crests (maximum displacement) or two troughs (maximum displacement). Wavelength is a measure of distance and hence has units of metres (m).

weight the gravitational force of attraction exerted by the earth on an object. As weight is a FORCE, its unit is the NEWTON (N). The weight of any object on earth can be calculated using:

$W = mg$ $m = $ mass (kg)

$g = $ gravitational constant $= 9.8 ms^{-2}$

In everyday use, the term weight really refers to the mass of a person or object.

Wheatstone bridge a divided electrical circuit used for measuring electrical RESISTANCE. The circuit comprises a diamond configuration of four resistances with a VOLTAGE applied between two points (between the points on the greater dimension of the diamond), and the opposite two points (the shorter dimension) bridged by a galvanometer (a device for detecting or measuring small currents). When no current flows through the galvanometer the resistances can be paired together to calculate one unknown, using a simple equation: $R1/R2 = R3/R4$.

whistler atmospheric electric noises producing whistles on a radio receiver. The effect is caused by lightning producing ELECTROMAGNETIC radiations, which follow the earth's magnetic field force lines and are then reflected back to earth by the upper atmosphere.

white blood cell *see* **leucocyte**.

white dwarf a type of star that is very dense with a low luminosity. They result from the explosion of stars that have used up their available hydrogen (*see* SUN). Due to their small size, their surface temperatures are high and appear white (*see* SUPERNOVA).

work an energy transfer that has the net result of mov-

work

ing an object. A FORCE is said to do work only when its object of application moves in the direction of the force. The work done is calculated by multiplying the force (F) by the distance(s) through which it moves, i.e. W = FS. Work is measured in the unit of energy, the JOULE. For example, if you have to pull with a force of 50 newtons to move a box 3 metres in the direction of the force (toward yourself), then work = 50N x 3.0M = 150Nm = 150J.

X Y Z

x-axis the "horizontal" axis in plane (two-dimensional) co-ordinate geometry.

X-chromosome one kind of CHROMOSOME that is involved in the sex determination of an individual. A woman has a pair of X-chromosomes, whereas a man has one X-chromosome and one Y-chromosome. There are many GENES on the X-chromosome which have nothing to do with the sex of the individual. For example, red-green colour blindness is determined by a RECESSIVE gene on the X-chromosome. If a woman carrying this gene has a son (X,Y) then he will inherit colour blindness as the Y-chromosome will have no corresponding gene to suppress the effect. If she has a daughter (X,X), and the father has normal vision, then the recessive gene is still inherited but its effect is suppressed as the X-chromosome from the father will carry the dominant gene for normal vision.

xerography a copying process in which an ELECTROSTATIC image is formed on a surface when exposed to an optical image. A powder mix of GRAPHITE and a thermoplastic resin of opposite charge to the electrostatic image is dusted on to the surface and the parti-

X-rays

cles cling to the charged areas. The image is then transferred to a sheet of paper, again through use of opposite charges, and the image is fixed by heat.

X-rays the part of the ELECTROMAGNETIC spectrum with a wavelength range of approximately 10^{-12} to 10^{-9} m and a frequency range of 10^{17} to 10^{21} Hz. X-rays are produced when electrons moving at high speed are absorbed by a target. The resultant waves will penetrate solids to varying degrees, dependent on the density of the solid. Hence X-rays of certain wavelengths will penetrate flesh, but not bone or other more dense materials. X-rays serve both therapeutic and diagnostic functions in medicine and are deployed in many areas of industry where inspection of hidden, inaccessible objects is necessary.

y-axis the "vertical" axis in plane (two-dimensional) co-ordinate geometry.

Y-chromosome the small chromosome that carries a dominant gene for maleness. All normal males have 22 matched pairs of chromosomes and one unmatched pair, one large X-chromosome and one small Y-chromosome. The X-chromosome, which is inherited from the mother, carries many more genes than the Y-chromosome, which is inherited from the father. During sexual reproduction, the mother must contribute one X-chromosome, but the father can contribute either an X or Y-chromosome. The effect of the Y-chromosome is that a testis develops in the embryo instead of an ovary. Thus the sex of the resulting offspring is dependent on the father's contribution—female (X,X) or male (X,Y).

yeast unicellular micro-organisms that form a fungus. Yeast cells can be circular or oval in shape and reproduce by spore formation. The enzymes secreted by yeasts are used in brewing and baking industries as they can convert sugars into alcohol and carbon dioxide.

yield point HOOKE'S LAW states that for a material such as steel, in wire form, the extension is proportional to the tension, up to the elastic limit. An increase in tension beyond this limit takes the material to the yield point, where a sudden increase in elongation occurs with only a small further increase in tension.

Young's modulus a method for calculating a ratio concerning the elasticity of a solid. Young's modulus (E) relates the STRESS and STRAIN in a solid (usually wire) using the following:

$$E = stress/strain = \sigma/\varepsilon$$

where σ = Force/Area and ε = Change in Length/Length Young's modulus has units of newtons per metre squared (Nm^{-2}) and is calculated only when the material is under elastic conditions, i.e. the applied force does not exceed the elastic limit and cause deformation.

zeolites a group of natural and synthetic hydrated alumina silicates of sodium, potassium, calcium and barium that contain loosely held water that can be removed by heating and regained by exposure to water. However, the cavities created by the loss of water can be occupied by other molecules of a similar size. Zeolites have thus found uses in ION EXCHANGE and as adsorbents. Zeolites containing small amounts of platinum or palladium are used as CATALYSTS in the CRACKING of HYDROCARBONS.

zoology a branch of biology that involves the study of animals. Subjects studied include anatomy, physiology, embryology, evolution, and the geographical distribution of animals.

zwitterion the predominant form of an AMINO ACID when surrounded by a neutral solution (pH 7). The structure of a zwitterion and amino acid differ in that the zwitterion exists as a dipolar ion—the carboxyl (-COOH) group of the amino acid loses a hydrogen atom to form -COO⁻, and the amino group (NH) gains a hydrogen atom to form -NH⁺.

zygote the cell immediately produced by the fusion of male and female germ cells (GAMETES) during the initial stage of FERTILIZATION. The zygote is a DIPLOID cell, formed by the fusion of the haploid male gamete and the haploid female gamete.

zymogen an inactive form of an ENZYME. Most zymogens are inactive precursors of pancreatic enzymes, which are involved in protein digestion. Synthesis of these digestive enzymes as zymogens prevents the unwanted digestion of the tissue in which the enzyme was made. The zymogen becomes activated by chemical modifications to form its functional form when it reaches its site of function, e.g. the enzyme chymotrypsin (digests protein) is synthesized in the pancreas as the zymogen, chymotrypsinogen, and becomes activated only when it reaches its destination, the small intestine.

zymurgy a branch of chemistry that involves the study of FERMENTATION processes.

APPENDIX 1

Periodic Table

Group	1A	2A	3B	4B	5B	6B	7B	8	8	8	1B	2B	3A	4A	5A	6A	7A	0
	H 1																	He 2
	Li 3	Be 4											B 5	C 6	N 7	O 8	F 9	Ne 10
	Na 11	Mg 12	←					TRANSITION ELEMENTS				→	Al 13	Si 14	P 15	S 16	Cl 17	Ar 18
	K 19	Ca 20	Sc 21	Ti 22	V 23	Cr 24	Mn 25	Fe 26	Co 27	Ni 28	Cu 29	Zn 30	Ga 31	Ge 32	As 33	Se 34	Br 35	Kr 36
	Rb 37	Sr 38	Y 39	Zr 40	Nb 41	Mo 42	Tc 43	Ru 44	Rh 45	Pd 46	Ag 47	Cd 48	In 49	Sn 50	Sb 51	Te 52	I 53	Xe 54
	Cs 55	Ba 56	La[1] 57	Hf 72	Ta 73	W 74	Re 75	Os 76	Ir 77	Pt 78	Au 79	Hg 80	Tl 81	Pb 82	Bi 83	Po 84	At 85	Rn 86
	Fr 87	Ra 88	Ac 89															

[1] Lanthanides	La 57	Ce 58	Pr 59	Nd 60	Pm 61	Sm 62	Eu 63	Gd 64	Tb 65	Dy 66	Ho 67	Er 68	Tm 69	Yb 70	Lu 71
[2] Actinides	Ac 89	Th 90	Pa 91	U 92	Np 93	Pu 94	Am 95	Cm 96	Bk 97	Cf 98	Es 99	Fm 100	Md 101	No 102	Lr 103

APPENDIX 2

Element Table

Element	Symbol	Atomic Number	Relative Atomic Mass*
Actinium	Ac	89	{227}
Aluminium	Al	13	26.9815
Americium	Am	95	{243}
Antimony	Sb	51	121.75
Argon	Ar	18	39.948
Arsenic	As	33	74.9216
Astatine	At	85	{210}
Barium	Ba	56	137.34
Berkelium	Bk	97	{247}
Beryllium	Be	4	9.0122
Bismuth	Bi	83	208.98
Boron	B	5	10.81
Bromine	Br	35	79.904
Cadmium	Cd	48	112.40
Caesium	Cs	55	132.905
Calcium	Ca	20	40.08
Californium	Cf	98	{251}
Carbon	C	6	12.011
Cerium	Ce	58	140.12
Chlorine	Cl	17	35.453
Chromium	Cr	24	51.996
Cobalt	Co	27	58.9332
Copper	Cu	29	63.546
Curium	Cm	96	{247}

Element	Symbol	Atomic Number	Relative Atomic Mass*
Dysprosium	Dy	66	162.50
Einsteinium	Es	99	{254}
Erbium	Er	68	167.26
Europium	Eu	63	151.96
Fermium	Fm	100	{257}
Fluorine	F	9	18.9984
Francium	Fr	87	{223}
Gadolinium	Gd	64	157.25
Gallium	Ga	31	69.72
Germanium	Ge	32	72.59
Gold	Au	79	196.967
Hafnium	Hf	72	178.49
Helium	He	2	4.0026
Holmium	Ho	67	164.930
Hydrogen	H	1	1.00797
Indium	In	49	1114.82
Iodine	I	53	126.9044
Iridium	Ir	77	192.2
Iron	Fe	26	55.847
Krypton	Kr	36	83.80
Lanthanum	La	57	138.91
Lawrencium	Lr	103	{257}
Lead	Pb	82	207.19
Lithium	Li	3	6.939
Lutetium	Lu	71	174.97
Magnesium	Mg	12	24.305
Manganese	Mn	25	54.938

Element	Symbol	Atomic Number	Relative Atomic Mass*
Mendelevium	Md	101	{258}
Mercury	Hg	80	200.59
Molybdenum	Mo	42	95.94
Neodymium	Nd	60	144.24
Neon	Ne	10	20.179
Neptunium	Np	93	{237}
Nickel	Ni	28	58.71
Niobium	Nb	41	92.906
Nitrogen	N	7	14.0067
Nobelium	No	102	{255}
Osmium	Os	76	190.2
Oxygen	O	8	15.9994
Palladium	Pd	46	106.4
Phosphorus	P	15	30.9738
Platinum	Pt	78	195.09
Plutonium	Pu	94	{244}
Polonium	Po	84	{209}
Potassium	K	19	39.102
Praseodymium	Pr	59	140.907
Promethium	Pm	61	{145}
Protactinium	Pa	91	{231}
Radium	Ra	88	{226}
Radon	Rn	86	{222}
Rhenium	Re	75	186.20
Rhodium	Rh	45	102.905
Rubidium	Rb	37	85.47
Ruthenium	Ru	44	101.07

Element	Symbol	Atomic Number	Relative Atomic Mass*
Samarium	Sm	62	150.35
Scandium	Sc	21	44.956
Selenium	Se	34	78.96
Silicon	Si	14	28.086
Silver	Ag	47	107.868
Sodium	Na	11	22.9898
Strontium	Sr	38	87.62
Sulphur	S	16	32.064
Tantalum	Ta	73	180.948
Technetium	Tc	43	{97}
Tellurium	Te	52	127.60
Terbium	Tb	65	158.924
Thallium	Tl	81	204.37
Thorium	Th	90	232.038
Thulium	Tm	69	168.934
Tin	Sn	50	118.69
Titanium	Ti	22	47.90
Tungsten	W	74	183.85
Uranium	U	92	238.03
Vanadium	V	23	50.942
Xenon	Xe	54	131.30
Ytterbium	Yb	70	173.04
Yttrium	Y	39	88.905
Zinc	Zn	30	65.37
Zirconium	Zr	40	91.22

*Values of the *Relative Atomic Mass* in brackets refer to the most stable, known, isotope.

APPENDIX 3

THE GREEK ALPHABET

Name	Capital	Lower Case	English Sound
alpha	Α	α	a
beta	Β	β	b
gamma	Γ	γ	g
delta	Δ	δ	d
epsilon	Ε	ε	e
zeta	Ζ	ζ	z
eta	Η	η	e
theta	Θ	θ	th
iota	Ι	ι	i
kappa	Κ	κ	k
lambda	Λ	λ	l
mu	Μ	μ	m
nu	Ν	ν	n
xi	Ξ	ξ	x
omicron	Ο	ο	o
pi	Π	π	p
rho	Ρ	ρ	r
sigma	Σ	σ	s
tau	Τ	τ	t
upsilon	Υ	υ	u
phi	Φ	φ	ph
chi	Χ	χ	kh
psi	Ψ	ψ	ps
omega	Ω	ω	o

APPENDIX 4

The International System of Units (SI units)

Quantity	Symbol	Unit	Symbols
acceleration	a	metres per second squared	ms^{-2} or m/s^2
area	A	square metre	$m2$
capacitance	C	farad	$F (1F = 1AsV^{-1})$
charge	Q	coulomb	$C (1C = 1As)$
current	I	ampere	A
density	ρ	kilograms per cubic metre	kgm^{-3} or kg/m^3
force	F	newton	$N (1N = 1 \, kg \, ms^{-2})$
frequency	f	hertz	$Hz (1Hz = 1s^{-1})$
length	l	metre	m
mass	m	kilogram	kg
potential difference	V	volt	$V (1V = 1JC^{-1}$ or $WA^{-1})$
power	P	watt	$W (1W = 1Js^{-1})$
resistance	R	ohm	$\Omega (1\Omega = 1VA^{-1})$
specific heat capacity	c	joules per kilogram kelvin	$Jkg^{-1}K^{-1}$
temperature	T	kelvin	L
time	t	second	s
volume	V	cubic metre	$m3$
velocity	v	metres per second	ms^{-1} or m/s
wavelength	λ	metre	m
work, energy	W, E	joule	$J (1J = 1Nm)$

APPENDIX 4 (cont.)

Useful prefixes adopted with SI units

Prefix	Symbol	Factor
tera	T	10^{12}
giga	G	10^{9}
mega	M	10^{6}
kilo	k	10^{3}
hecto	h	10^{2}
deda	da	10^{1}
deci	d	10^{-1}
centi	c	10^{-2}
milli	m	10^{-3}
micro	μ	10^{-6}
nano	n	10^{-9}
pico	p	10^{-12}
femto	f	10^{-15}
atto	a	10^{-18}

APPENDIX 5

Geological Time Scale

Eon	Era	Sub-era	Period	Epoch	Millions of years since the start
PHANEROZOIC	Cenozoic	Quaternary	Pleistogene	Holocene	0.01
				Pleistocene	2.0
		Tertiary	Neogene	Pliocene	5.1
				Miocene	24.6
			Palaeogene	Oligocene	38
				Eocene	55
				Palaeocene	65
	Mesozoic		Cretaceous		144
			Jurassic		213
			Triassic		248

APPENDIX 5 (cont.)

Geological Time Scale

Eon	Era	Sub-era	Period	Epoch	Millions of years since the start
PHANEROZOIC	Palaeozoic	Upper Palaeozoic	Permian		286
			Carboniferous		360
			Devonian		408
		Lower Palaeozoic	Silurian		438
			Ordovician		505
			Cambrian		590
PROTEROZOIC PRE					2500
ARCHAEAN CAMB					4000
PRISCOAN RIAN					4600

287

APPENDIX 6

The Solar System

Planet	Diameter at the Equator km	Mass relative to the Earth[1]	Average distance from Sun km⁶	The planet's "year"
Mercury	44840	0.054	57.91	87.969 days
Venus	12300	0.8150	108.21	224.701 days
Earth	12756	1.000	149.60	365.256 days
Mars	6790	0.107	227.94	686.980 days
Jupiter	142700	317.89	778.34	11.86 years
Saturn	120800	95.14	1427.01	29.46 years
Uranus	50800	14.52	2869.6	84.0 years
Neptune	48600	17.46	4496.7	164.8 years
Pluto	3500	0.1 (approx)	5907	248.4 years
Sun	1392000	332 958		
Moon	3476	0.0123		

[1] The mass of the Earth is 5.976×10^{24} kg